On Breathing

On Breathing

CARE IN A TIME OF CATASTROPHE

Jamieson Webster

CATAPULT  NEW YORK

First Catapult edition: 2025

ISBN: 978-1-64622-241-4

Library of Congress Control Number: 2024943549

Jacket design by Nicole Caputo
Jacket image © Boris Austin / Millenium Images, UK

Catapult
New York, NY
books.catapult.co

Printed in the United States of America

10 9 8 7 6 5 4 3 2 1

I, who had experienced so many pleasant diversions, stood on the bank and watched. "What are our lungs supposed to do?" I shouted. Shouted: "If they breathe fast they suffocate themselves from inner poisons; if they breathe slowly they suffocate from unbreathable air, from outraged things. But if they try to search for their own rhythm they perish from the mere search."

—Franz Kafka, "Description of a Struggle"

I am as confident as I am of anything that, in myself, the stream of thinking (which I recognize emphatically as a phenomenon) is only a careless name for what, when scrutinized, reveals itself to consist chiefly of the stream of my breathing. The "I think" which Kant said must be able to accompany all my objects is the "I breathe" which actually does accompany them.

—William James,
"Does 'Consciousness' Exist?"

It is unreasonable to be asked not to breathe. But it is miraculous to breathe for another.

—Anahid Nersessian,
Keats's Odes: A Lover's Discourse

On
Breathing

HAVING ANOTHER CHILD brought me back to an intimacy with air and breath that I had forgotten. A whole world of bubbles and balloons and breathing sounds, from sniffling to snoring to snorting pigs. The breath that animates the bubble is also the breath behind the sheer power of a voice as it expresses joy or rage or agony. The bubble vanishes in the same way that sound does. Or, I have been reminded, a breath can hang around encapsulated by a balloon.

The magic of blowing a kiss feels like invisibly crossing a vast distance, and air can be pushed against skin to make the most delicious sounds and sensations. We are bathed in breath with others' speech. And long before they have learned to form breath into words, the infant enjoys manipulating air to form rhythmic sounds. What does any of this air say?

While adults tend to marvel at that first act of breathing independence by the newly born infant, it is the closeness to breathing throughout childhood—in fact, throughout life—that interests me, along with the ways the delicacy of breath can make it a magnet for anxiety or a target of violence. From the moment of birth, breathing is something we must do for as long as we are alive, yet for the most part we don't pay it much notice. If one of the main claims of psychoanalysis is that we forget sexuality, I think we must add breathing to our list of amnesias.

In fact we are *meant* to forget that we are breathing. I remember a patient, love forlorn, who obsessionally tracked her breathing against the beating of her heart, driving herself half-mad in the process. The biological cadence of breath is forced into the background, and I imagine this distance from consciousness is what allows us to override this rhythm in such extraordinary ways. Do the glass blower, the blues singer, and the imitator of the mating calls of birds feel an intimacy with ways they use breath?

Think back to the early pleasures of the mouth and its intake of air. This is not the exquisite sensibility of the gourmand, the pleasant sting of that first whiskey for the stressed-out worker. And it's not the same as the oral erotic pleasures of adult sexual life. It's something closer to the sheer aliveness of the child, cold air manipulated at the back of the throat, tongue tapping on your teeth while sucking in, water allowed to dribble down your chin. It's a high-pitched scream in an underpass with its wafting echoes reverberating on the walls. It's whistling.

My dear friend, the psychoanalyst and artist Bracha L. Ettinger, writes about the many psychoanalytic theories that see desire as the key inheritance we accord to our children—what we convey that we want for ourselves, for them, from each other, and from life more generally. But desire, she points out, is not the same as aliveness. A feeling for life must be transmitted to the child beyond any given desire; it is part of giving birth to them. *Hospitality, welcoming, ingathering, carrying* are the words she uses to depict this primary "matrix" of care.

I always think of the videos of toddlers from the psychoanalyst Margaret Mahler's 1973 study of separation. When the toddler first begins to walk independently, they return again and again to the mother, joyously filling her lap with objects from their travels. Mahler called this phase the "love affair with the world." What is so striking is that the affair doesn't last long, ushering in a terrifying period of ambivalence about this newfound independence. Little by little, learning to speak ameliorates the crisis. Through speech, the infant will learn to articulate desires. Speaking is the breathing that carries you back into your love affair with life.

No other movie has sparked such passion in my daughter as *The Red Balloon*. A friendship between a child and a balloon inevitably confronts the fragility of all balloons, whose fate is flying away, popping, or deflating. In the end, a thousand balloons will rush to the scene, saving the child, whose beloved companion was cruelly taken down by a mob of mischievous children. But such serenity facing loss cannot merely be the result of the love object, this balloon, being revealed as a fungible object. Surely there is something else?

In the film's final acts, the bereft child is carried by balloons through the skies above Paris. Is this an image of exaltation for the balloon as more air than object? Is the extension of oneself as breath into a balloon what counts, meaning we need only imagine ourselves being lifted into the air to rejoin our lost friend? "There is a solidarity between the soap bubble and its blower that excludes the rest of the world," writes the philosopher Peter Sloterdijk. "For its creator, the soap bubble thus becomes the medium of a surprising soul expansion."

The idea of solidarity with the air feels paradoxical. One can't help but hear Marx's infamous pronouncement on the reach of capitalism: "All that is solid melts into air." We have now toxified our air. Somewhere, we must feel unending climate grief—what writer Cade Diehm says we middle children of human history face: "too late for the earth, too soon for the stars."

As pockets of asphyxiation are increasingly finding their way into life, twisting a basic and primary pleasure in breath into new forms of violence, I try to remember this ephemeral extension of ourselves. Sloterdijk named the lacunae around air "negative air conditioning," reversing soul expansion with high-power vacuum suction. My daughter—born shortly after I was working with those dying of COVID as a palliative care psychoanalyst—returns me to a breath that isn't death.

One day a balloon suddenly popped; horrified, my daughter didn't want another one around for quite some time. She was devoted to balloons, and until that moment, she seemed practically fearless. How had her most passionate object of love become her first source of fear? It was the sound that haunted her. She didn't want to hear it pop and seemed willing to sacrifice this intimate pleasure to prevent a newly experienced pain from recurring.

At the time, she was navigating potty training, leaving behind some of the final markers of infancy. As with toddlers, she was also confronting an onslaught of frustrated adult voices and their infinite prohibitions. I had the fantasy—maybe the psychoanalytic intuition—that the fearful pop of the balloon was a translation of both the feeling of losing control in her body (a body that she was newly working so hard to know and master) *and* an experience of those booming, angry adult voices. I watched her transfer her love of balloons to a dream of flying through the air (as a fruit bat that she called "fruit bag") for the time it took her fear to evaporate.

This little story of our love affair with air contains my hope for this book. That the memory and thought of breath gives us some courage in dark times, reminds us of our capacity to offer care to something delicate that traverses our lives. To attend again to something we have forgotten. Isn't that always the task of psychoanalysis—to pry open our amnesias?

Through respiration, as through psychoanalysis, we address the link between body and soul. It's worth remembering that the Greek word *psyche* (like the Latin *spiritus*) means both "breath" and "mind," so that, even if its practitioners have forgotten it, breath has always been at the very center of psychoanalysis. Thus there is always the possibility of returning to the fugitive, but primary enjoyment of everything we can do with breath in a body—until our last.

FIRST
BREATH

FOUR PUSHES AND my daughter was out. A first breath. A little cry. And she was delivered into my arms. I didn't scream much, or do any quick-paced Lamaze breathing like they do in the movies. The scene was quiet. She looked on at me intensely.

My son—born almost ten pounds, nineteen years earlier—paved the way for her quick birth. He flipped his car on the drive down to the hospital and was miraculously untouched though the car was totaled. After he was given the okay in the emergency room, he made his way down to the hospital to greet his new sister, recounting how he'd undone his seatbelt and crawled through the crushed windshield. He was . . . born again? I try not to interpret the accident, even though I am a psychoanalyst. One morning and two births, two harrowing passages into life, is what I'll say about it—with a deep sigh of relief.

The lungs are the last organ a fetus develops. During pregnancy they're filled with amniotic fluid. The first breath is usually taken within ten seconds of delivery, though some babies require help clearing mucus from their noses and mouths, or seem to wait for that little pat on the back, which I can't help but think is performative more than necessary—as if we really need an authority to slap us into life. The infant—now in the world—will have to breathe as long as they are alive.

The theorist Leo Bersani sees in the newborn's first autonomous gulp of air an assertion of independence and existence. An assertion that comes all on its own:

> Breathing is the tiny human's first experience of her body's inescapable receptivity, a taking in which is inseparable from a letting out. Breathing initiates the dual rhythm of receptivity: absorption and expulsion. Repeated continuously and involuntarily through-out human life (we become aware of it only when it is momentarily blocked), it is the most fundamental model of the organic dualism intrinsic to all animal life.

For many parents the moment of waiting to hear that first cry—the sign of life, breath, arrival—will have been imagined many times before it happens. How often, as a young child, I played at giving birth, the appearance of the screaming infant, the doctor's authoritative words. I loved to lie at the bottom of a swimming pool, looking at the light at the surface, and rush up toward it desperate for breath. The dual rhythm of breathing is always in the background of life, but ready to make its overt appearance in those aspects of life that speak life—in eating, in sex, in physical exertion, in anxiety, and of course in speaking itself.

To be "born" means to be carried and sustained. It refers, oddly, to the work of the mother rather than what takes place for the child. The mother's body sustains a baby until they can take in something foreign that will then sustain them—air. Birth, as the experiencing of coming to breathe, is traumatic for the infant, and so an environment of parental care is necessary for quite a long time. Air carries the infant's cries, and these cries call on the parent. The air that moves from the world into the interior of the body also carries the sounds that reach our ears. Millions of years of evolution for this powerful little feedback loop.

If breathing is the sign of our separation from the bodies of our mothers, our first independent act, then it is also able to speak to these relations of care. In her memoir *Fierce Attachments*, Vivian Gornick writes about her intimacy with her mother as if it is something that contaminates air: "Her influence clung, membrane-like, to my nostrils, my eyelids, my open mouth. I drew her into me with every breath I took. I drowsed in her etherizing atmosphere." The British pediatrician and psychoanalyst Donald Winnicott named maternal care of the infant "the holding environment"—not only because an infant must be held constantly, but also because parental care must provide a container for the infant's anxiety. If the love is too little, the infant feels abandoned. If it is too much, the infant feels smothered. In Winnicott's terms mothering needs to be "good enough," the middle term between the extremes of presence and absence.

It's true that the psychoanalysts have always posited that dependency, something that begins in relation to the mother's body, is a psychic threat throughout our life—as if we could be swallowed up again, as if we hadn't fully separated. So, we breathe and know more intimately our dependency, or we breathe as an act of defiance against influence. Separation is a fact, but it is also a fragile process. As with breathing, there are vulnerabilities, forms of resistance, ways of co-opting what seems, at least on the surface, as though it should just *happen*. Perhaps this is one reason we take breathing for granted, this invisible dependency and shared reality that we cover over with an illusion of independence or self-reliance.

I like to imagine my daughter hearing herself for the first time along with her first breath. I imagine this memory of hearing herself after a timeless time of being unable to make sounds in utero. She lets out a first gasp with a cry—*whaaaaaaaa*. I had already thought about the sound of her voice as she moved inside me, responding to the sounds of my voice. I also imagined her breathing, watching the rise and fall of her chest as she was sleeping. This is a world of rhythmicity that solicits. Hearing her cries. Rocking her to sleep. Watching her breathe.

Freud believed our first memories are centered on the sound of our own crying. Our cries are indelibly etched into our minds alongside whatever experiences of pain or fear as well as the soothing by others that (hopefully) follow. All memories have an acoustic accompaniment that goes back to these first ones—a double archive in the mind. Freud uses the German *schreien*, which means crying out, bawling, wailing, screeching, yelling, and which in the standard edition of his works has been translated into English as "screaming." We are, in the Freudian universe, utterly helpless as human infants. And yet, the infant has this power to solicit.

The modulation of breath into a cry is a tool of survival that is also the beginning of memory—one that stretches all the way back to the beginning of the species, maybe even life. Eventually our memories of crying develop into the wish to communicate, and then into the ability to do so. Through language we are more able to get the help we need, though we only master this skill *after* we most needed it. (Strange fact of being human: this prolonged helplessness, this disjunction between need and the acquisition of skills.)

Psychoanalysis believes that we return to some part of this archaic and repressed memory when we are with our own infants, and one might speculate this is the reason they can induce such powerful emotions in us, certainly love, but also experiences of deep anxiety or depression. I remember when my partner slyly smiled at me as I tried to explain to him the reasons for our daughter's every whimper during those first few weeks home. I am, after all, a psychoanalyst. I should know something about this.

I should also know that the desire that someone know is one of the most enduring and powerful desires—and that, like all desires, it is born from a lack. My partner seemed to know better. With gentle sarcasm, he asked me if this was my leading theory for the moment. It was the leading theory. For the last five minutes. My explanations finally exploded on me one day when, after insisting that she was crying because she had colic (or excessive gassiness), I read online that no one knows what colic is and that it may be that babies have colic because they cry—meaning they swallow a lot of air. Why are they crying then? Because that's just what babies do? How had I just travelled in a complete circle? Were we both so besieged by helplessness?

I felt totally mad. Also mad at the fact of how little we understand about infants. At the very least I just needed to return to being with her. Winnicott called this state of mind "primary maternal preoccupation," which sounds nicer than it is, since he also called it a necessary madness or psychosis. The idea is that you become so completely enveloped in the world of the infant that you do become a bit crazy, but this madness is helping you provide an environment as close to the womb as possible. A womb-like protective environment despite breathing and crying and internet searches that lead nowhere.

In the kind of slightly offensive anachronism the internet likes to propagate, this period is now called "the fourth trimester." The baby cries, sleeps, and eats, and we must live in this rhythmicity twenty-four hours a day, keeping watch and not demanding too much or letting the world intrude. Perhaps one does need to go a bit mad to tolerate all the crying. Freud speculates that the foundational memory of crying in every one of us is the starting point for empathy, but empathy, in the world of psychoanalysis, knows no bounds and can become a total merger—thus madness. Fear of the infant's helplessness can lead to smothering. Or abandonment. The point isn't to not be afraid, however, but in fact to make room for it.

The subtle process of differentiation and separation begins with birth, but it does not end there. Winnicott and many other psychoanalysts speculate that the infant does not know the difference between themselves and the other. They exist in a state of imaginary symbiosis. But very quickly, they begin marking difference. Separation from the body of the mother in birth, the loss of the intense relationship to the oral object when beginning to eat, and the acquisition of language—these are the key moments in a differentiation that is never quite finished and where progress is always precarious.

When speaking we don't hear our own voice, a subtle loss that is always at play. This capacity to screen out our own voice breaks down in psychosis—one's voice returns from the outside as an alien entity that torments. It is a return to a state of helplessness worse than that of the infant, because no one can help. Only heavy medication that blunts everything, destroys the rhythmicity of life. One of my first patients on the psychiatric ward used to desperately ask me to go to the "lungsbank" to get him lungs; he claimed he didn't have any and thought it would help him with all the screaming voices in his head. I thought there was something brilliant in this psychotic thought. He wanted new lungs: the engine of breath, speech, and so life itself. He wanted to be born again.

My daughter came to have a very serious pair of lungs on her. She is strong, stronger than me and my son—able to bring everyone into her orbit and demand what she wants with voracity. While pregnant, I was giving endless lectures and I sometimes imagined she had to listen to me talk, talk, talk. As she began to show her prowess with sound, I fantasized that she was responding in kind to my torturing her in utero. But her manner in general is outward facing, directed to the world. See me! Hear me!

She devoured my breasts and milk as if they were her total possession, sometimes squeezing and slapping them for more. I watch as she inhales the smells of my perfumes, letting out an *ahhhhhhh* with gusto, only to throw the bottle down in glee. Or the way she relishes what she eats with a demonstrable sigh and a fistful of food, a sardonic smile at those looking on a bit awkwardly at this state of consumption. Her first word was *that*. First words are called holophrases. They compact many strands of meaning and complex ideas into one utterance. *That*, a word of pure exhalation, was designed to bring the world to her for her consumption. All these pleasures of the mouth and tongue and lungs and air are bound together.

Freud might link the development of memory and language to crying, but he also links the organization of the mind to pleasure. The pleasures of mouth are first—Freud's oral stage. The psychologist René Spitz said the baby "sees with its mouth." Inhale everything with your mouth. This isn't just a literal definition of what it means to breathe. Everything that leans on breathing in the world of the child brings pleasure—eating, drinking, sucking, tasting, smelling, and yes, babbling, yelling, singing, and even screaming.

The power of an infant's suck is magnificent. Feeling it on your nipple or finger is like a first taste of an inborn life force that seems to expand forever outward—life straight into life. We meet the infant here. This world of pleasure balances the scales of so much helplessness. Infant, meaning the one who cannot speak for herself yet but who can use her breath in all sorts of other ways. I wish I could breathe with you from your first breath to your last.

AN AMERICAN PILOT named Dr. Forrest Bird, who had worked on aeronautical oxygen regulation for flying at high altitudes during World War II, was the first person to make a reliable, low-cost, mass-produced ventilator. He modelled the 1958 machine on the way air moves across the wings of a plane: airfoils, he realized, were like alveoli in the lungs. A brilliant transposition of organ to machine. His device made iron lungs obsolete, and he eventually created the Baby Bird, which radically reduced the mortality rate of infants born prematurely.

Dr. Bird's preoccupation with breathing followed him into his own life. His inventions failed to prevent the deaths either of his first wife, who died of lung damage after severe bronchitis, or of his second wife, who died in a plane crash. But his Baby Bird did save his stepdaughter, who was born a month premature. (She now runs the museum that bears his name—the Bird Aviation Museum and Invention Center in Hayden, Idaho.)

However, the most significant advances in medical ventilation have taken place in the last twenty years, when computers have begun to address the problems caused by machines forcing air into delicate lungs, often injuring them in the process. It took a long time for doctors and scientists to understand that in the long run you cannot force someone to breathe.

The new machines can finally sense the diaphragm's spasms that signal the desire to breathe, and the breath's timing and increasing and decreasing volume of intake, and then match both to a tee. This mimicry of the human body by machines could be a saving grace, especially in the face of an inability to modulate human interventions on life more generally—as a species we really cannot gauge what is too much or not enough. Modern technology is always a double-edged sword, contributing to the destruction of the planet and people that it is then called on to save.

Psychoanalytic technology is old. Psychoanalysis is about speaking—say anything, say everything—and I've come to see breathing as a hidden navel in our field. Speaking requires the modulation of breath. Freud famously called the psychoanalyst a midwife, but the question of being born, birthing a subject, becomes even more interesting when this act is seen as an ongoing one of breathing and being in the world. How can breathing and being in the world combine with speaking as freely as possible? What will this free speech bring to a question of being, especially being with others?

Breathing is founded on a separation from the body of the mother, and so on a separation from others more generally that is always incomplete. But it is also important as a sense for the bodily rhythmicity of life and a new dependence on air—an air we all share. Breathing is the space where extreme anxiety, the mysteries of the body, and the scars of being born rear their head. Breathing eventually transmutes into language, and the act of speaking is crucial in this story of breath, where being born seems to need to happen, at the very least, twice: first biologically, as a human mammal, and second as a social animal that breathes in the air with others, and whose breathing can be impacted by them.

For many years, I had heard of a mythical creature, a psychoanalyst who had a position in a neonatal intensive care unit at a hospital in Paris. This analyst would whisper to the babies their history, the conditions of their coming to be born, and who their parents were, in an effort to speak them into life. She had the audacity to ask the question as to why some babies survive and not others, why some succumb to sudden infant death syndrome or failure to thrive, and to suggest that the answer was far more mysterious and complicated than medical pathology could account for. She decided it was a psychoanalytic problem.

Dr. Catherine Vanier wasn't always such a grand figure. As she tells it, she was treated as someone like a janitor and a witch. When she first began in the hospital, she was even given a broom closet for her first office. She was, as she put it, "tolerated" by her colleagues until they began to see the value of her work. It seems important that she was also available to the staff, to help with their anxiety in the face of such tiny humans.

Thanks to better ultrasound technology, we know how much is heard and responded to in utero. Vanier points out that even as we began to be able to save premature infants—"preemies"—it took time for us to see them as having been born and in need of being treated accordingly. Until about thirty years ago, doctors thought that premature infants didn't feel pain, so they operated without anaesthesia. These infants were enclosed in boxes and hooked up to machines and feeding tubes, devoid of human contact. This made weaning the infants from these devices more difficult. Who was being protected in this scenario?

Protocols began to change fifteen years ago, recognizing the importance of contact time between newborns and parents, though the explanations for the new procedures were purely behavioural—the skin-to-skin produced oxytocin. Really? Why do these explanations stop short of addressing the infant as someone born, a true breathing, living subject, even if a nascent one? Why reduce the infant to biochemical receptors? Just think of the pomp and circumstance around births dubbed *on time*. The clothes, the naming, the blankets, the balloons. Wheeling the baby around in manic glee. Behold the new child!

They put a ridiculous giant bow on my daughter's head, something I had a hard time accepting. One could speculate that a great deal of this ritual is for the parents, to help accommodate them to their new life, not only the new life of the child, but also their new status as parents, even if for a second or third or whatever time. We could speculate that what was left out in the situation

with preemies is the reality of the parents as parents, given that they are so terrified that their infants might die at any moment. What is being managed is what Vanier often noticed—that the parents wanted to abandon their infants to the care of doctors, not ready to accept the reality of having a new child, especially one so helpless and small. So, we closed the space of anxiety by acting as if there wasn't a child there.

Skin-to-skin then was helping the parent bond to these tiny babies, to be less afraid of them, and therefore to make room for them in their psyche. To take them in. And this was in turn showing the child that there was someone, beyond a machine, to form an attachment with. Vanier tells the story of a baby named Anna, who would immediately go into respiratory distress every time doctors tried to take her off the respirator. One day, the care team saw that she could be taken off the ventilator if the ventilator was left running so its sound was audible from her incubator (it had been mistakenly left on once it was removed). If the sound stopped, she stopped breathing. It was as if her respiration could only go on if minimally propped up by the rhythmic noise of the whirring machine. Isn't this incredible?

Vanier was summoned to find an answer to the strange question of what could wean her off this need for the sound of the machine. Vanier speculated that it could be the voice of the mother, who had not spoken much with her daughter. Vanier encouraged her to speak to her child often, showing the mother that the baby reacted strongly to the sound of her mother's voice. It was also important, Vanier insisted, that the mother speak the truth—her worries, her anger, her concern, her love, all of it; empty speech would not do.

Speech that is full is speech that addresses the infant, sees the child as having been born, as living, as having a future that the other imagines and desires. Otherwise the speech would be empty of affect, of an ability to really impact Anna's body—to call her into life. She made this vital transfer from the machine to mother, showing how the mother's words impact the breathing organ of the baby, which will eventually be the organ that allows Anna, one day, to speak to her mother. Think of this conduit, lung to lung!

I find this so beautiful and right. And yet I must admit that reading Vanier's work, I wrestled with a perverse cynicism: she imagines she's speaking these infants into life and breath but really it's her imagination. This was a strange sentiment since I, too, am a psychoanalyst, and I have experienced the quasi-magical power of speaking, simply speaking, to affect the body. I also believe in speaking in the direction of truth as best one can. Why would I want to deny something so fundamental?

Was it resistance to touching this primordial point, laying open this scene of utter helplessness, one I had lived in my own childhood, the infancies of my children, and on many occasions with my patients? In fact, like Vanier, I was a psychoanalyst in unlikely places of extreme questions of life and death. I jumped into the line of fire after the 9/11 attacks, working with firemen and policemen and rescue workers exhausted from searching for bodies in the rubble. I was a psychoanalyst in the hospital during the first wave of COVID. How could I be cynical?

I had witnessed what it means to intervene as a psychoanalyst at the moment when life hangs in the balance. Donning a mask, did I just want to imagine that I was heroically working in aberrant emergency circumstances and not confronting my own utter helplessness? I do think it is hard to acknowledge this most basic element of life—breathing. And to know that we may be born, but we are not all carried to term.

Sunny Day
Sweepin' the clouds away
On my way to where the air is sweet
Can you tell me how to get?
How to get to Sesame Street

FOR THOSE OF us who grew up on *Sesame Street*, its opening song is a gateway to the atmospheric memory of being a small child. Small, as if to follow the song's subtle alliteration in the sound of those drawn-out *s's—sunny, sweeping, sweet*. *Sesame Street* was the utopian creation of Joan Ganz Cooney and Lloyd Morrisett in 1969. I first watched the show in the early 1980s, and then again with my daughter not long after working in palliative care during COVID. Forty years on, the air no longer promised a neutral, nourishing environment. It carried and circulated the plague. It had been contaminated and made toxic by modern industrial civilization.

In the contemporary imagination, "sweet" has been taken over by images of sickness, scarcity, suffocation, and smothering. The snake eats its tail: the climate crisis means being unable, at some point, to breathe—whether the image is one of smoke, excess water, toxic air, unbearable temperatures, or simply the cessation of breath that is death. Thinking about this can induce panic, whose main symptom is hyperventilation. We can't breathe imagining that we can't breathe.

I saw this tight claustrophobic loop of anxiety on the palliative ward. I watched as staff, utterly overwhelmed by fear, grew violent in the face of patients struggling to breathe. I don't blame them. The situation was impossible. The monstrous anxiety of faltering breath, carrying an incurable disease, was as contagious as COVID—spreading from patient to doctor. In those first months, intubating a patient and having a machine breathe for them was the only treatment. During those first months I helped family members speak to their loved ones via iPads before they were put into a medical coma for ventilation, knowing it could be the last time.

We only later understood that intubation often diminished a patient's chance of survival. Patients, in fact, could continue to breathe despite very low oxygen levels and a feeling of constraint—something many pulmonologists were attesting to. But this would require staff helping the patients to tolerate anxiety about not being able to breathe, while also tolerating their own fears of infection, suffocation, and death. Here were staff who were completely burnt-out on death—not just of patients but also

fellow essential workers. Still, in this reliance on intubation, I can't help but wonder, who was being protected?

The rush to procure ventilation machines was likewise misguided, a siren song for politicians to wage war
on one another and feel like they were being helpful by
stockpiling breathing machines. Fear and helplessness
seemed to be dictating so many decisions. I think we
were overly comforted by the idea of a machine stepping
in, forgetting the necessity of mutual care we needed to
offer, a failure that also led to the massive mental health
crisis following the isolation of lockdowns. As I worked,
the staff continued to threaten patients with intubation
when their own anxiety became overwhelming.

I watched patients in the ICU develop psychosis,
either as a response to being in the ICU—alone, on ventilation machines, heavily medicated and with no sense
of time or place—or as one of the strange side effects of
COVID. One patient, who was delirious, was tied to her
bed, wrists and ankles, like a psychiatric patient, because
she kept getting up and pulling out her oxygen tubes. The
staff didn't want to assign someone to sit in the room
with her "one-to-one," as they call it. I eventually quit vol-
unteering because I couldn't watch this cycle any longer.
My helplessness was turning into rage.

THE SCENES I witnessed in the hospital brought me back to when my oldest child developed asthma at the age of two. There was desperation as he pressed his small body into the floor of the emergency room, as if trying to ground himself. Probably he was trying to open his diaphragm, the reason we bend forward when we feel breathless. But my own experiences with asthma as an older child shot into memory: the funny feeling that being upright and not being able to breathe was insulting. "If I can't breathe in this air, if I can't assume the posture that makes me human, leave me to the ground," was how I heard my thoughts.

Granted this is a dramatic, even hysterical response to breathlessness, but I felt like my dignity was being taken from me in having to struggle for something so elementary. This thought is probably a condensation of years of struggling that became a certain attitude or composure. As Céleste Albaret wrote of Proust's asthma, "He looked very young . . . And then his exquisite elegance and that peculiar manner, a kind of restraint, which I later noticed in many asthmatics, as though he were husbanding his strength and his breath."

The exaggeration betrays something real that must have been a part of the development of asthma. Asthma has long been seen as sometimes having psychosomatic correlates, especially in infancy and childhood. Saying this doesn't mean that the asthma isn't real, nor that the effects of pollution or other environmental stressors aren't real. In fact, the problem with anything psychosomatic, as psychoanalysis conceives of it, is that the real effects in the body take on a life of their own. One must treat the problem biologically *and* psychologically.

I developed asthma at the age of five after my mother left for the Philippines to go to medical school. Like Dr. Bird, my father was a pilot and was often away on trips, so I was left on my own with a nanny. Somehow I managed to have asthma most often when my father *was* around. The psychoanalyst Otto Fenichel felt asthma, stuttering, and tics were about the neurotic involvement in natural functions. He heard a longing for maternal protection, a wish for soothing of these bodily spasms.

Freud's student Sándor Ferenczi speculates that certain cases of asthma and epilepsy are caused by a failure in the caring environment to bring them to life, to push back against what he calls non-being. "Slipping into non-being," Ferenczi writes, "come[s] much more easily to children," since the "life force" that protects them is not yet strong enough. Making a child feel loved, secure, and wanted can help "bind" them to life—love as an immunization against physical and psychic injury. As with many asthmatic children, my asthma disappeared with the onset of adolescence, a fact that cannot really be accounted for by the biological maturations of puberty. Perhaps the separation from childhood is key.

Years later my son developed asthma while his father and I were splitting up. I remember how my son would scan us, looking at our demeanor towards him, our shared anxiety, as much as we were monitoring his difficulty with breathing. He would often reassure us—"I'm okay, Mommy!" Probably, in the reversed logic of all unconscious messages, this is a message to us: Please be okay! Together in the emergency room it felt like one of the rare times we were united in our care for him. In this scene, breath, for the first time since being in utero, was a palpably shared experience.

ONE OF THE strangest memories I reckoned with in psychoanalysis was being turned upside down, face down, in a tilting hospital bed while a vibrating device was applied to my lungs twice a day for weeks. It is in fact a real treatment, but it seemed medieval to me as a child, or, from another angle, a silly parlor game. It comes close to some of those turn-of-the-century *Magic Mountain* cures via water baths or disguised orgasm devices—the very ones that I studied when I investigated the history of treatments for so-called hysterical women. "Vibration is life," reads a 1910 advertisement headline. "Vibrate your body and make it well."

The treatment did make me well. The vibration and extreme positioning loosened the mucous built up in my lungs from repetitive infections leading to chronic bronchitis and asthma. This strange therapy was helping to clear my airways. They were sweepin' the clouds away. I liked the attention I was receiving, the ceremonial preparation for these treatments after long hours in a hospital.

I remember some embarrassing erotic feelings towards one of the therapists. At least I think I remember them. In psychoanalysis, erotic transferences can lean on parental ones during moments of extreme vulnerability. The open backs of those hospital gowns while upside down!

Behind this memory lay another series of memories of childhood games I hadn't thought about since they happened. We used to play a fainting game by compressing each other's necks, causing a momentary blackout. I believed, in accordance with my asthmatic universe, or perhaps the cinematic image of strangulation, that we were cutting off our airways. In fact, we were cutting off circulation in the carotid artery that carries blood and oxygen to the brain. Back against the wall, we would black out, slide down onto the ground, slumped over, and then wake up. It took no longer than a minute. Certainly dangerous— or so the internet tells me now—but at the time it felt perfectly innocent and thrilling. Leaving the world and coming back to it felt miraculous. A vast journey.

Another memory: many girls during a sleepover, in the dark, chanting, "Concentrate, concentrate, babies are crying, people are dying, concentrate, concentrate." Various gestures are performed on one girl, her head, her neck and torso, from pounding her back and declaring that she has been stabbed and the blood is running down, to an egg being cracked on her head and the yolk running down her hair and neck. Towards the end she is pulled with an imaginary rope attached to four corners on her back, to which she magically complies, or many hands slide under her body, which gives the feeling of levitation. Sometimes she stands and is tussled while being whispered to that she is dizzy and might fall off the ledge of a building, giving her a feeling of vertigo. Sometimes, after a large clap of the hands, the lights turn on, and the colour immediately seen is a prophesy of one's death—yellow is poison, grey is disease, purple is suffocation, and so on.

All of this is so macabre in retrospect, but at the time it was a real thrill. A whole world of new sensations in the body. Having the wind knocked out of me when I fell forward off my bicycle, hitting the handlebars, then the ground. There I was, in some stranger's front yard on the side of the road, gasping for air. It must have been one of the few times I was far from home, on my own, forced to handle the emergency alone. I watched my breath return to me and promptly rode home to tell someone (who?) all about it. That's when I was told, "Oh, you got the wind knocked out of you." I didn't know what was happening to me. I have wind?

My psychoanalyst asked me where the parents were in all these memories, as if to say that there was something more to it, something a little unreal. One of the few times you were on your own, on your bike, was when this happened? Handlebars, handle yourself? Yes, why not that one time! But where are the parents in all these scenes? Where are they when seven children are making each other faint in a kitchen? Where are my parents during these vibration treatments?

These are prepuberty memories and games. They are games before the major new sensations of sexuality funneled in, when some are oddly cured of childhood asthma. The memory of being alone, or being only with friends, this experimentation with the body on the border of anxiety and excitement. Adolescence is anticipated, if not investigated, by these games. One is excited or terrified alone, or excited and terrified with strangers or peers. It is not a familial scene even when it harkens back to them. When my asthma disappeared around fourteen it felt like a miracle. I simply didn't have it anymore. I didn't need an inhaler. I didn't develop bronchitis every year. I even started smoking cigarettes with some trepidation— though defiance was the overriding attitude. I loved smoking. I still do. The rebellion was certainly directed at my parents, but it must also have been directed at the asthma, trying to overcome being the child who couldn't breathe. It also makes me wonder about the fear that is missing: fear of dying, fear of being alone, fear of not being a child anymore.

"IT WOULD BE interesting to discuss the relative importance of the control of the bronchus as an organ," wrote Winnicott in 1941, in his paper "The Observation of Infants in a Set Situation." At the Paddington Green Children's Hospital, where he worked for twenty years, Winnicott created a method for observing caregivers and their babies. In that essay, he turns to the case of a seven-month-old who had been suffering from bouts of asthma, as did her mother as a child and during pregnancy. He was able to observe one such attack, and wrote of it:

> The breathing out might have been felt by the baby to be dangerous if linked to a dangerous idea—for instance, an idea of reaching in to take . . . It will be seen that the notion of a dangerous breath or of a dangerous breathing or of a dangerous breathing organ leads us once more to the infant's fantasies.

Asthma, Winnicott says, is an involuntary control of expiration, a difficulty in letting go of one's breath, or, from another angle, an excessive holding of breath. It can only be a fantasy of imagined repercussions, he posits, that would stop the baby from surrendering to a natural impulse. When natural inclinations are brought to a halt unnecessarily, symptoms erupt. Fortunately, this kind of inhibition in infants, in contrast to that in adults, can be reversed quite quickly by allowing the wish to carry on in its merry way. Babies are so new to the world, miraculously open to help.

Meeting the infant's desire for help is what was a "good-enough," in Winnicott's words, "facilitating environment." The way he sees the secure infant in his observational situation is fascinating. Three temporal moments transpire, which all have a logical, rhythmic structure, one that almost feels likes breathing. The composition is uncannily close to the three temporalities that the French psychoanalyst Jacques Lacan described, using the prisoner's dilemma from game theory: the instant of the glance, the time for understanding, and the moment for concluding.

To begin, the mother—in Winnicott's essay, it is only mothers, characteristically of the time—enters and walks across a large room with her baby to Winnicott, who is sitting at a table. She is asked to sit catty-corner across the table from him. The scene of arrival is important so that the mother, child, and doctor have a chance to see one another and make visual contact before settling in. Winnicott then places a shiny, right-angled tongue depressor (which he calls a spatula) on the table. He tells the mother to sit in such a way that if the baby should wish to take the object, it can—but she is instructed not to help or interfere and to contribute as little as possible to the situation. The classical setting with couch, patient, speech, and analyst is here transformed into mother, baby, spatula, and doctor.

During what Winnicott calls stage one, the baby puts its hand out to the spatula but realizes that the situation requires more thought. The baby grows still but not necessarily rigid—looking at the doctor or its mother with big eyes, or withdrawing from the situation and burying itself in the mother's chest. No reassurance is given, since what is crucial is the spontaneous and gradual return of the child's interest in the seductive object. Like the moment of the child's glance, what dominates is a question about what this object, and this strange other, means for the baby. Who are you? Who am I? What is possible here?

This leads to the next stage, in which the hesitation conflicts with desire. The baby must become "brave enough to let his feelings develop," which will change the encounter quickly, an alteration evident first in the child's body. The baby's mouth becomes "flabby," its tongue "thick and soft," and it might even begin drooling. Stillness then gives way to movement with a decisive action. With the spatula finally in hand, the infant displays a sense of confidence, even power, giving rise to play—putting the spatula in its mouth, joyfully banging it on the table, offering it to its mother's mouth, or the doctor's, pretending to feed them like a baby. "See" with your mouth, test the world with this, your first device.

One might imagine that this would be the final stage, and Winnicott himself says he thought so too—that it took him some time to see there is an important final move. In stage three, the infant drops the spatula, perhaps by accident, but then plays with getting rid of it, which is thoroughly enjoyed. The infant may want to get down and mouth the object again on the floor, but will eventually lose interest in the spatula, leave it behind, and turn its attention to the wider room. Leaving the scene of the object is a second act of bravery.

The infant must feel that what is left can be returned to, and yet this can never be guaranteed. If the moment for concluding an encounter is always the negotiation of an exit, with the baby this comes after the triumphant exploration of its satisfaction. The baby cannot completely understand, or square the loss implied in leaving, but must leave nevertheless. This means risking giving up what has been found satisfying in order to look further afield and step into an unknown future.

So what happened with the asthmatic seven-month-old? Paralyzed in the time of hesitation, the baby may have felt too greedy, too wary of retribution, too great a threat of loss, or without hope for more satisfaction. Basic anxieties must be worked and reworked with help from the outside. This isn't about the perfect protection of parents that may, in fact, make one more neurotic; rather, the complexity of the infant's feelings is held by a good-enough caretaker until they can unfold in their own time, becoming more distinct desires, contained by a parent. Could we say that the asthmatic infant is thrown back into its being, especially that of being a body? The asthmatic is inhibited in the face of the object that they cannot take, play with, or feel untrammeled pleasure with.

The infant at stage two, on the other hand, creates an object, is seduced, plays at being seductive, turning subject into object, passive into active. In the game of feeding mother just as one is fed, the infant has a self-representation on the outside to test, perhaps a little like curating one's image on Instagram and looking to see how many people like it. Winnicott even points out that since he is there, the infant has the object by negotiating two presences, two onlookers, and must have a certain amount of courage.

Only after the glittering object is ready at hand can the infant explore not-having, losing it, dispensing with it—a breach that opens onto the wider world. This makes me think of when my son was able to give up an obsessive concern with the integrity of his Lego figures, which were finally, to my great relief, exploded all over the room.

We now have three distinct temporal waves—being, having, not-having. Beginning and end meet; being is close to not-having, but it is modified by the step of having-had. One might even say that being is finally transformed through an act of mourning, of loving and losing, rather than being a place of retreat, of losing oneself entirely, or sticking with the repetitive power of having.

Traversing these three moments, Winnicott says, is a preparation for the advent of language. We are thus born thrice—into the world as a being, a being who can have and be had by others, and as someone who can then have the courage to not-have, to give some things up and look for others, to speak about what is not immediately there.

WE OFTEN FORGET that a fetus breathes using the placenta, so caught are we in the image of the baby like a fish. As a child I was an avid snorkeler; later I would become a deep-sea diver. Of course the asthmatic becomes a diver! The best divers know how to conserve their oxygen use to stay under longer, barely using their extremities for momentum. Just breath. The average person can hold their breath somewhere between thirty and ninety seconds. If 100 percent oxygen is breathed, the world record is twenty-four minutes and thirty-seven seconds.

Breathing might seem natural, an autonomic function taking place beyond consciousness, but it can be controlled in quite extreme fashions, and is certainly available for involvement in neurotic symptoms. In fact, many of the first patients in psychoanalysis experienced choking symptoms, difficulty swallowing, feelings of suffocation, small tic-like coughs or *tussis nervosa*. While many think of paralytic limbs and emotional or epileptic fits when picturing Freudian hysteria, there really is a background of symptoms around breathing. Freud gave an example of what we may call respiratory erogeneity in 1895: "A woman suffered from attacks [of] obsessional brooding and speculating . . . The theme of her worry was always a part or function of her body; for example, respiration: 'Why must I breathe? Suppose I didn't want to breathe?' etc." Freud's relentless cigar smoking should perhaps also be included as a symptom, part of the Victorian tobacco trance.

Suppose I didn't want to! Enter the world record holder for breath holding. The ability to hold one's breath underwater is a learned skill. Studies suggest that we can tap into an evolutionary reflex so that, instead of triggering involuntary breath, we activate a vestige of aquatic life from billions of years ago. In his book *Thalassa: A Theory of Genitality*, the psychoanalyst Sándor Ferenczi saw leaving the sea and crawling onto land to breathe air and engage in sexed reproduction as a catastrophe still present in our mental life. Maybe it wasn't about not wanting to breath but a wish to return to the sea.

The sea for Ferenczi was an all-encompassing, nourishing, fluid environment where we were bisexual or transsexual and fertilization took place outside in the watery expanse. Once on land, the wish to go home takes the form of semen, "the little fish," returning to the womb. The sea is introjected inside in the form of amniotic fluid. Our crashing upon the shore is repeated in the shock of birth. Sexuality, for Ferenczi, is a regression to our primeval history as both oceanic bliss and the survival of the threat of desiccation during prior moments of climate catastrophe.

Genitalia and reproductive organs differentiate and evolve on land, since air—unlike water—is unable to carry and mix gametes. This mixing becomes a property both internal to and between bodies. Ferenczi feels that the death drive is present in our extreme individuation. Especially the forced domination of one half of the species for the purposes of reproduction inside the body of the other half. Human sexuality develops as a process of friction and penetration. He sees this as an evolutionary aberration where breathing and human sexuality are the vulnerable bodily sites that will one day be subject to extreme threat again.

I find the Thalassa theory so moving. We want to go back home to the ocean. We want less friction, this bumping of bodies against one another, private property and the acquisition of land, the war between the sexes. Isn't so much of what is taking place today an attempt to return to a more fluid sexuality? To undo, or at least reckon with, the violence and catastrophe that has resulted from life on land? This catastrophe for Ferenczi is the beginning of anxiety, and so it surely makes sense that one of the few examples of mastering anxiety is the ability to hold one's breath and swim underwater again.

IN HIS SHORT story "Café Loup," Ben Lerner writes about his morbid fear of his daughter choking. Compulsively imagining her choking becomes prophetic, not of her choking, but *him*—on a bite of steak caught in his throat while having dinner with a friend. Struggling to breathe, Lerner fears that he will die a ridiculous death that falls short of the vainglorious suicide that he has meticulously and neurotically planned for after *his daughter's* accidental choking. The fate he was trying to avoid, he has brought down on himself:

> I have always thought there was sadism haunting the question "Are you choking?," since the one who still draws breath and can form it into speech demands from the person who is choking a response they cannot give . . . choking is a uniquely human drama, a definitional drama for the homo loquens, a drama at the heart of the human, or, rather, at the larynx, the voice box, which, as we evolved, moved lower and lower, enabling

us to generate a long column of vibrating air we can shape into meaning with our mouthparts, shaped air that might in turn build and shape a world, but this evolutionary "speech advantage" required that the space in our bodies for breathing and swallowing be shared; the formal capacity for speech comes with the risk of choking to death, something only humans frequently do.

Choking, like language, is a human affair: our highest and most fragile gift is, as Lerner suggests, this column of air that we can manipulate. Lerner speculates that psychoanalysis, his Jewish religion, might interpret that he has brought this act of choking upon himself as punishment for tempting fate with all his imagining of his daughter's premature death. So, still struggling for breath, he makes a promise to the gods that if he's allowed to live, he will go the way of silence, contemplation, noncompulsive imagination of choking others. He may even quit being a writer. Stop all this coughing up of words: "I will learn the way of silence, I will no longer manically ingest and express, will neither tempt fate nor attempt to evade it with talk."

Clearly, he lives to tell the tale. And in *The New Yorker*. It's simply too good of a story—maybe the story of all stories—so he's almost guaranteed not to choke putting it down on paper.

There is a moment in Lerner's story where he says that people choking often report extreme shame and hesitate before signaling they need help. Many want to rush out of the scene and handle the situation themselves. Why this need to control even one's own choking? Lerner almost wants to follow the airline emergency protocol that asks us to help ourselves before our children. And yet in a kind of reversal of that logic, it is as though he has choked first to better assist his child: Lerner learns from this episode to finally stop tormenting her, chopping up her food into tiny bits, vigilantly watching her every gulp, and micromanaging her mother.

If choking is a uniquely human drama, it is because of the way self-consciousness can interfere with the natural ability to regulate air intake. Every year of elementary school we had to run the mile for the presidential fitness challenge, which was finally abandoned after over fifty years in 2013. I hated running. I felt weak and always had to stop. I didn't do well with tests in general. I didn't like seeing where I lined up. I remember clutching my inhaler in my pocket and stopping to use it—something I never did on any other occasion. I didn't have the kind of asthma that suddenly manifests. I didn't have asthma attacks.

There was something performative about my asthmatic struggle on the track around the soccer field. Bending over, acting like it was such an effort to breathe deeply. Mind you, south Florida is hot, really hot. A stark contrast to the cold puff of medicine delivered to your throat and lungs with an inhaler. I hated the way the medicine made me feel jittery, and I had to be careful not to let this performance get the best of me. All the other kids were there waiting for me to finish my mile. This act of using the inhaler seemed to offset the feelings of self-consciousness that I was always last or second-to-last in the "challenge." Smoking cigarettes served me similarly, putting distance between me and others, the reassuring feeling of having something ready to hand.

There is a test that was given to measure lung capacity—the spirometer—where you breathe out into a tube and a ball or marker moves a certain length indicating a number. It felt on the one hand so scientific, so discreet in its score-giving mechanism. Yet again, it was also like a children's game. A party blower! I never knew if I was faking it or not, or in which direction. I failed to show how compromised I was by trying too hard to breathe out—so macho! Or I failed to breathe out to my full capacity, wanting to make myself look deficient for some gain. Malingerer. But what did I want to gain?

The French psychoanalyst Jacques Lacan remarked that the psychotic person has the object in their pocket, and others are seen as wanting to steal it from them. This object must always be had. The psychotic person senses the other's interest, even enjoyment, very closely, and thus easily feels intruded upon. The neurotic—or so called "normal"—person feels it is the other that has the object that fascinates them and that they lack. Their neurosis takes the form of symptoms and interpersonal struggles as a game of demanding this object or pretending to ignore it to get the other to proffer it. A neurotic always wants to win by way of cheating, squaring the odds, levelling difference, covering over this feeling of lack that motivates them.

Is the inhaler the object in my pocket? Or a neurotic game around a difficult feeling of lack? My parents called it my "puffer." I always thought of the fish that blew itself up like a balloon, a fascinating evolutionary trick that developed, according to some marine biologists, from a family of fish that coughs and another family of fish that blows jets of water. The puffer fish gulps water and expands one hundred times its own size, scaring away predators but also creating a hard shell, should their size fail to deter. Over the course of evolution, this capacity to gulp water made puffer fish lose a functional stomach. They use their digestive tract and liver for the same purposes. Gain some, lose some.

But this loss of a stomach actually made the fish hardier, better able to survive in different concentrations of salinity, eat just about everything, and live amid pollution. Many scientists think they should be studied for their capacity to protect themselves in hostile environments, having given up their stomachs. Some puffer fish attract mates by creating beautiful circular patterns on the ocean floor that have been compared to crop circles. The mesmerizing grooves in the sand also protect their eggs by changing the pattern of the flow of water around them. These fish are really artisans of breath.

In her work on "autistic breathing," the British psychoanalyst Maria Rhode writes about children who manipulate breath. Rhode points out that, while breathing and eating are united in utero, in life they are distinct processes that can interfere with one another. She theorizes that some of her patients treat breath as food, meaning they use it to feel full, providing a form of control less dangerous than eating, which, in childhood, always involves another person.

A patient called Charles would hyperventilate and use noisy breathing to—as Rhode interprets it—seal himself off from others completely. He would breathe in sync with rhythmic rocking or tapping as a means of self-soothing, but also as a way of creating a kind of sensorium, a protective environment, around himself. Rhode surmises that this seemed to arise at moments in which he felt the analyst's boundaries, and thus felt more out of control, separate from her. Speaking to him about these feelings enabled him to loosen his grip on his breath. In a later session, he once punched himself in the stomach, winding himself, which she interpreted as Charles showing her how separation felt to him like suffocation or a loss of breath.

Another patient who came to treatment with Rhode at the age of four was obsessed with balloons, which Rhode thought of as breath made visible and kept within. The string on the balloon symbolized for her the umbilical cord—the tie to the mother. In session, this patient would react to a girl he seemed to think was the analyst's daughter and begin filling his cheeks with air like big balloons. This obsession with balloons and air, through treatment, was eventually transformed into singing to himself, which allowed him to dare to really speak his first words. Rhode says this was an important turning point.

Psychoanalytic categories aren't always helpful, and children and adolescents are, in any case, exempt from diagnostic categories. Rhode is unique in treating autistic patients and even seeing meaning in their symptoms, a way they can unfold in a process that has the potential to cure them and to bring them closer to dialogue with others. For Rhode, what these children exhibit are beautifully crafted symptoms composed of breath.

Looking back on my own asthma, I can identify with Rhode's patients, though I was not adept with my breath the way they are. In college I loved the feeling of having a pack of cigarettes close at hand, similar to its predecessor, the puffer. There was a certain pleasure when my boyfriend gave me the pet name Half Lung—a reference to how, huffing and puffing on the stairs, I'd complain I have half the lung capacity of other people. I have less! Give me something!

I probably received a diagnosis during childhood spirometer tests, but it was certainly more a position I was holding. A way of condensing anxiety into a single symptom. In my later twenties I was finally able to test my lungs athletically and through speaking. I like the idea that the affection of this name was what I needed, and that my college boyfriend was intrinsic to me finding my voice. This is also the job of the psychoanalyst. Not the giving of pet names, but a work of naming that meets the patient halfway, aerating suffering and gradually evaporating it at its most dense.

ANXIETY

WHEN ANXIETY DESCENDS it thrusts breathing into the foreground, revealing a closed claustrophobic loop in life: anxiety causes a constriction of breath, and the constriction of breath increases anxiety. In the throes of an anxiety attack it can almost feel as though one has forgotten how to breathe. A recent study of high school students concluded that anxiety and depression were worse among those who took mindfulness classes than among those who didn't: as if drawing attention to anxiety simply risks breeding more anxiety and trying to learn to breathe was subject to too much resistance.

The place of anxiety in any therapeutic cure is much more complicated than the easy alliance between the anxiety that characterizes childhood and the work of parenting that seeks to ameliorate it. Freud saw traps everywhere: the parent's excessive love for the child could exacerbate its anxiety, leading to an adult anxiety that is much more difficult to contain. Children should therefore face the unpleasantness of life free from the affectionate fantasies of adults. Childhood anxiety made children appear like neurotic adults while it turned neurotic adults into children. What do we do with adult anxiety? How do we understand it?

When he was starting out, Freud would try to hypnotize patients by pressing his hands to their foreheads, but in the psychoanalytic method he went on to develop, touch became ferociously forbidden. Since then it is as though there is an invisible line separating patient and analyst, as though the fundamental rule of psychoanalysis—say everything, no matter how impolite or sexual, no matter how anxious it makes you—was born out of this prohibition of touch.

Clearly, I have a confession to make. I once massaged a patient's arms during a panic attack. Her hands started to stiffen, the fingers coming together and curling over upon themselves, like the long, rounded claws of a sloth. The sight of this monstrous takeover of her body seemed to drive her anxiety into a full-blown panic attack. She looked practically possessed. Hyperventilating, she pleaded that I tell her what was happening to her.

"Don't look at your hands," I said, running over to Marilyn, "look at me." She suddenly sat upright on the couch, held my gaze, then sat on her hands, pulling them from under herself to try and open them. I instinctively moved to help her, massaging her arms and hands, trying to gently open her fingers. The muscles on her arms were taught. Rigid.

She was still breathing hard, but I went to work trying to make her body relax. Eventually her hands loosened, bringing the temperature down in the room. I found it strange that I attended to Marilyn's body first and not her state of mind, given what I do. But it was seeing her own hands that produced her anxiety and inability to breathe.

This wasn't the first time I'd noticed her hands. I used to look at her incredibly long acrylic nails while she was lying on the couch. I wondered about those claw-like nails, how or whether they incapacitate you. I also admired the rebellion they seemed to imply. Form over function. My hands don't belong to you. More figuratively, I often thought about her difficulty with getting her hands around life.

Panic was a new symptom in a case characterized by generalized anxiety and alexithymia, or an inability to experience or name emotions. Previously a dancer, everything seemed to be expressed in her body. (Or on it—she had full-body tattoos, neck to feet.) A few months before the scene in question, she told me about a panic attack while driving to confront a lover. Getting closer to the border between New York and Massachusetts where he lived, she was suddenly overcome by anxiety.

Her heart was racing, and she started to feel a sensation of suffocation. She had to pull the car over. She couldn't hold the wheel; her hands were numb. Eventually the police came, an ambulance, but she managed to calm down and drive herself home. She never made it across the state line.

This second panic attack happened during the penultimate session of a long analysis, just after she graduated from college and right before she moved out of town. She had been describing packing up her apartment. Exhausted, she'd drunk a little too much. She admitted, finally, to being sad, maybe anxious. She had gone to the roof of her building for a cigarette and fell down the stairs—a repetition of something that had happened in her first years in psychoanalysis when she fell down a flight of stairs at a stadium.

What if the fall had been more severe? she'd wondered. After all, she was alone. No one would have been there to help. She needed to be more careful, especially now that she was leaving for a new city where she truly would be on her own. She hadn't really let herself think about this, the separation from those who looked after her, including the separation from me. As she spoke, she started to breathe quickly. Anxiety came over her. I felt it.

What is it that makes anxiety so palpable once it has entered the room? Anxiety is a breathing problem, to be sure. But it is also a problem of air that is shared. We unconsciously synchronize our breath with those around us, but one need not have an outward, physical manifestation of anxiety for it to begin spreading. Are we unconsciously cued to other, less perceptible signs? How can the other's anxiety so fearsomely disturb my own contentment? How does anxiety thrust what is in the background into the foreground?

Once I was able to open Marilyn's hands, I asked if she felt like we needed to take her to the hospital. She said no, she was okay, and immediately called a friend. This flight from intensity felt typical, even as she had just met me with the immensity of her anxiety. I said emphatically to call me if "it" came back. I also noted that she kept this for the second-to-last session, meaning we had one more chance to meet again, to talk about what *was* happening.

After Marilyn left, I thought about her sitting on her hands to undo the spasm. I heard the phrase "sitting on your hands," implying a passivity in life, not extending your hand to the other, to not offer help—waiting indefinitely. I thought about how I had to jump out of my relatively passive position as analyst and help her. I crossed a boundary I don't cross easily. I literally used my hands to open her hands. I thought about the long hiatus she had been caught in, financially dependent on her family again while she tried to finish college as an adult. I thought about how fraught their help was—they never extended a hand to her easily or openly.

We had wrestled with the way her desire to finish school thrust her into an anxious ambiguity about receiving help. What is help? She felt like a child. Taking opened what she had wanted to close. Taking for psychoanalysts is always a radical opening, acknowledging what one doesn't have—but wants. No wonder this paralysis of hands!

It felt nearly impossible to form a picture of Marilyn's early life, shrouded in obscurity the way anxiety blanketed her. She was a runaway as a teenager, living counter to the outward values of her upbringing. Did her family disapprove of her? Then why had they agreed to pay her way while she finished school? Did they want to help? Was it even help, given they had a lot they only shared with her reticently? Should she be grateful? Guilty? Angry?

These were painful questions that tore into her. She tried to approach the line where she might find out more, only to be met with blank stares. She would sink into anxiety not knowing what *anyone* thought of her—not her teachers, not her boyfriends, certainly not me—terrified of her image of their image of her. A hall of mirrors, these images were thin, blurred, distorted, menacing. You would almost need full-body armor or a decoy self to manage such an enemy. Anxiety and ambiguity are the culprits who walk hand in hand.

For Freud, anxiety moves from having no object—a state of immanence that is also the threat of total annihilation—to having an object that concentrates fear on the loss of it. Only then can anxiety blossom into fear of the loss of love, with all the ideal images of recognition or disapproval that file in. Hence the utility of breathing exercises: focusing on one's breath is a way to have the most minimal object, between nothing at all and all the stereotypical images that disappoint or harass.

Lacan noted that anxiety detests movement, especially the moment of *seeing* movement. Once in full movement, anxiety might abate. It's the transitions that are impossible. Lines appeared everywhere toward the end of Marilyn's treatment; lines that, like the state line at which she'd stopped, had been impossible to draw or move across. She was on the verge of graduating from college, graduating from psychoanalysis—dispensing with this position of anxious dependency. The word *graduation* comes from the Latin word *gradus*, meaning "to step," "to go." She was having trouble going, moving beyond college, her relationship, New York City.

Above all I think Marilyn was drawing a line on a childhood state of passivity that had come back to haunt her. This also meant drawing a line on an image of control born from that childhood. Anxiety makes impossible demands, while attempting to control the responses given, which always feel like they dominate, overwhelm, and dissatisfy. This is why so many psychoanalysts point to the sharp-edged aggressiveness of the anxious.

Marilyn taught me that separation anxiety doesn't always look like being too attached to someone—about the ways that anxiety wraps life in a veil of passivity and ambiguity. The image of agency rises like a promise over a horizon that is always receding into the distance like the vanishing point of a picture plane. Anxiety is the fixed perspective.

Whatever might shift it, especially the kaleidoscope of personal meanings, is forcefully eschewed. Anxiety is its own prerogative—resisting all transformation. The anxious compulsively need to find, see, or feel something that will assuage them, which sometimes amounts to pain itself. The domination of the idea of mind over matter; matter, the body, always wins in the battle with the anxious mind.

Anxiety's recalcitrance frustrates therapists to no end. The concerns of the anxious patient seem just too concrete, reality-based, shorn of the world of fantasy, desire, and imagination. They have very little autonomy while dreaming about a phantasmatic one that is absolute. Maybe this is why many therapists have claimed that the anxious should just focus on body work, meditation, and breathing techniques as they seek to gain a sense of control. These help, but ultimately, I don't think anxiety should only be quelled.

I credit Marilyn with showing me the possibility of the careful transformation of anxiety through the panic that concentrates anxiety *and* the emergence of the bodily symptom that represents the anxiety and calls out. Seeing her hands suddenly entering strange paralytic postures on the couch was like seeing one of those first psychoanalytic patients, the grand hysterics. Their amazing bodily contractures, epileptic fits, losses of consciousness, and bursts of memory followed no known medical knowledge. They left the psychiatrists stupefied. The sanatorium was like a medical side show. What kind of a body was *this*? The emergence of this freakish body was the beginning sign of real change.

It took Freud's discipline to stop being fascinated by the feats of hysterical patients—to go from looking or touching to listening. No one was listening. It's still a problem today. When Freud listened, he realized that their symptoms spoke to conflicts with their milieux, rebellions against the limitations of their lives, especially as women, and feelings of desire and sexual arousal that had no place. Hysterics were trying to acknowledge so many unacknowledged feelings, memories, pains, sexual experiences good and bad. Hysterical symptoms were a genuine transformation of anxiety, substituting the anxious body for another one, more fantastical, more rebellious, more alive. A body to have a conversation with. An embodiment.

Being a dancer must have been Marilyn's first attempt to move in the direction of another kind of body and a different sense of being in the world. It worked for some time, especially as an adolescent and into her early twenties. The difficulty of sustained love and desire in relationships, along with her obscure desire to finish school, required something new of her. I was there to take up the running the second time around. She returned to dancing close to the end of our work.

As for her wish to study, Marilyn couldn't say much about it. She implied feelings of inadequacy without explicitly saying so. She was referred to me by her psychology teacher, who saw that she needed help but also felt she had a lot of potential—more than she could acknowledge. Marilyn was, she told me, an excellent writer. Darian Leader points out that writing is the final embodiment of thought through the hands. And it is also an embodied memory of all the speaking Marilyn did not do. Speaking that is also a transformation of breath, a breathing through her hands.

We psychoanalysts don't like giving cases fairy-tale endings, but so far, this kind of panic never occurred again after the episode in my office. We spoke six years after her final session. After reading something I wrote about the pandemic, Marilyn reached out to me. In the session, she told me how happy she was with her new job and a new partner. For them, the pandemic hadn't been all that difficult. That's all she really wanted to say to me. Beyond this concision, I could see something on her face. More life, some room to breathe.

I HAVE NEVER had a panic attack. There are only a few incidences of extreme anxiety as an adult that I can remember, usually related to separation from a loved one. The feeling was certainly one of worry, a need to immediately start attempting to control reality, but the feeling of true panic never blossomed. There were no heart palpitations or worsening constriction of breath. I could hover shy of the edge where anxiety and your body greet one another as childhood friends.

If in spatial terms anxiety is fear without an object, then it is also, from the perspective of time, a deferral of panic. Anxiety stretches panic into an infinite future. With panic all the anxiety is finally there, all at once. This is why panic attacks can be a welcome sign. It means anxiety is trying to find its crescendo, its bodily and mental apex. Breaking in on itself like a cutter, only then can anxiety seemingly transform outwardly.

This transformation often takes the form of a phobia, meaning the fear begins to have an object. Phobia is the more developed form of anxiety and panic. Psychoanalysis has long recognized phobia as being claustro-agoraphobic, sandwiched between the fear of tight spaces, being buried alive, and the fear of being outside among the people, falling forever in the boundless universe. Going out and staying in are the Janus-faced problem of anxiety, and between these two highly symbolic poles of anxiety are the full range of minor phobias.

A list of 140 such fears was catalogued in the nineteenth century. We all know people with specific phobias, from fear of flying, fear of driving, fear of heights, fear of water, fear of public speaking, fear of throwing up in public, to fear of small animals (bugs, mice, worms, etc.), or fear of large animals (horses, lions, bears, snakes, and, interestingly, only sometimes fear of domesticated animals like dogs and cats). Fear of illness, hypochondria, and specific fears in relation to other people, be it fear of sex or some form of violence, are in a category of their own because, unlike the other phobias (and more like claustro-agoraphobia), they are less circumscribed.

Sándor Ferenczi quoted a patient describing her suffocating feeling of anxiety and hypochondria as like being a fly caught in a cobweb, or a baby born *en caul*:

> Hypochondria surrounds my soul like a fine mist, or rather like a cobweb, just as a fungus covers the swamp. I have the feeling as though I were sticking in a bog, as though I had to stretch out my head so as to be able to breathe. I want to tear the cobweb, to tear it. But no, I can't do it! The web is fastened somewhere—the props would have to be pulled out on which it hangs. If that

can't be done, one would have slowly to work one's way through the net in order to get air. Man surely is not here to be veiled in such a cobweb, suffocated, and robbed of the light of the sun.

Phobia takes anxiety, amorphous and omnipresent, wrapping you in its veil, and creates boundaries. Phobia reorganizes the web, leaving bigger spaces in which to breathe. Better to fear the spider than the whole web. If a spider is potentially nearby, anxiety! If a spider really is present, panic attack! But no spider: all is well.

I find the development of a generalized feeling of angst into its more specific, but less specified, forms fascinating. Illness and the ambiguity of another person's intentions, for example, are always there, an unknown we must live with. Likewise, feeling crowded by people or places, or feeling the vastness of the world or world population, is readable into any situation. What psychoanalysts have done so well is to show that these fears are not about the external world per se, but something in the heart of the anxious person. Did the person fearing illness feel something inside of them was sick? Did they have a death wish towards someone? Did the person who felt easily crowded want to run away from someone specific? Or did they want to invade someone passionately or violently?

For Freud, phobias allow for an easier navigation of life. Key to his theory was that the object had to be symbolic. Yeah, yeah, we all know, a snake is a penis. But in a way, one might not want to analyze this snake phobia too much since, as psychoanalysis teaches us, it is functioning

as a ballast, holding the entire system together, opening the non-snake world to the person in question. The psychoanalytic cure is not about simply extricating the phobia, but about enlarging this symbolic sphere, creating more space.

The question of anxiety is also always tied to the issue of breathing. Breathwork is certainly aiming at a relaxation of the overactive nervous system. Take, for example, the experience of claustrophobia, with its constriction of the breath, the tightness in the chest, the feeling of being locked into one's body. Suffocating, smothered. Death, of course, means that we take a final breath. But claustrophobia ties death directly to the taking away of our ability to breathe, in comparison to other deaths of which the last breath is secondary to some other bodily emergency. As psychoanalyst Otto Rank noted:

> The fearful idea of death as a scythe-bearer, who severs one sharply from life, is to be traced back to the primal anxiety which man reproduces for the last time in the last trauma, in the last breath at death, and so gains from the greatest anxiety, namely, that of death, the pleasure of denying death by again undergoing the birth anxiety.

How paradoxical is Rank's image of anxiety. Anxiety is secretly the pleasure of the denial of death through the assertion of always being reborn to anxiety.

Is this the secret fulcrum of the infinite expanse of anxiety—death-denying expansion? I think he could be right. Being buried alive, for example, to use the most iconic expression of claustrophobia, brings death and last breath as close as possible, literally into the grave: our womb tomb. A cozy image, I suppose, a return to an airless space.

Agoraphobia takes the opposite track, not an inability to breathe but the opening of our bodies towards the world as air in a confrontation that speaks to the smallness of our body in comparison to the atmosphere, the universe, space. Sometimes described as the feeling of being poised above the abyss, the sensation that one is falling forever, it is the other side of claustrophobia. Air is the first foreign element that we take into our bodies, and having separated from the uterine environment, we must begin to learn to live in it—a lifelong task. Air is also the air we share with all those other people. Who hasn't, like the agoraphobic, dreamed of taking what is foreign and shared and making it into their own possession—locking it away in their house along with themselves?

FREUD BELIEVED THAT in repression something unacceptable is pushed out of conscious awareness and into the unconscious. For Freud repression functions by leaning on a more archaic repression that he called primal or primary repression. Repression was merely a false bottom under which there was an even greater unknown. The place where we encounter this false bottom he called the navel, as if to suggest that the greater unknown was always at root the maternal body, from which we have never fully separated ourselves.

This "beyond" has attracted the psychoanalysts' passion to no end. And constantly it seems to be bound up with the work of breathing. For a century, psychoanalysis has been fascinated by the trauma of birth, by the infant's first breath, which is controlled not only by biology but also constitutes a first experience of anxiety. Could we unlock the problem of anxiety by reaching down into this abyss? Was this the locus of the soul? Can we meet something greater than ourselves at this navel, like a world-collective unconscious? Is it the true reservoir of our libidinal store-house, a site not only for ecstasy but the sacred?

The first circle of psychoanalysts wanted to lay claim to this greater unknown, as if to undo repression in its entirety and thus to clear the way for a full, uninhib-ited breath, free from anxiety. Freud himself ultimately steered clear of this goal. It was all too mystical for him. "I suggest we quit such an unproductive field of inquiry without delay," he writes in 1926. Freud thought his dis-ciples were laboring under a fantasy of immanence, that this constituted a denial of the fundamental reality of anxiety, separation, and limit in a way that was not dis-similar to the promise of an afterlife or the sensation of eternal life and a "oneness" with the world that he named "the Oceanic feeling."

Anxiety is Freud's bugbear—he thought that if any-thing was going to be the ruin of society it would be anx-iety, which had the power to cause physical illness, mass hysteria, and mental disintegration. Freud was not inter-ested in undoing primal repression but in working with it. This meant working in air, however difficult. Rejecting oceanic fantasies in favor of the world of desire that blos-somed from repression meant helping his patients to learn to breathe less anxiously precisely by accepting dis-comfort and the lack of satisfaction that life entails.

Freud's more cautious approach to anxiety becomes clearest in relation to hysteria. It was women hysterics who taught Freud about anxiety precisely because they were incapable of experiencing it. Freud called the hys-teric's lack of anxiety her "beautiful indifference." He was fascinated by how the hysterical patient would erupt in extreme emotional reactions, bodily symptoms of all kinds, but magically never angst.

With a flick of her wrist, the hysteric dismisses the very possibility of concern, as if lassitude was her ultimate master, not the male analyst. And yet, paradox of paradoxes, she will suddenly take off running headfirst towards some traumatic scene and faint, as though she were moved by a strange will to repeat this anxiety that never manifests as such. For the hysteric, anxiety is no longer the conscious master but now the secret puppet master behind the iron curtain. And this, yes, the hysteric has no knowledge of, no sense of what this repressed anxiety is. But she could—through psychoanalysis. Getting in touch with the repressed would give her an access point to a whole scene unseen.

Importantly, this access route is not about undoing repression. Freud didn't want to dissolve this barrier, which creates and sustains a living space. Instead he wanted to secure it, or to resecure it. What is made conscious is never everything; only a portion will become known. The unconscious, as Lacan noted, will always remain Other. Breathing, I think, should be included here. It will also remain foreign ground. One can work with breath, but we will never master it in our lifetime.

This is why Freud had so much praise for hysteria and its modes of sublimation, which he saw as a symbolic defeat of anxiety. In the hysteric's symptoms we see how anxiety can be put towards the creation of something new, how we can learn to tolerate the unpleasant feelings of not knowing and being exposed in the sublimating creative act.

The importance of breath is the way it always intertwines self, body, and world. Thus sublimation is not an ecstatic release, an uninhibited intake of breath. It is a careful and long negotiation that manages to put the unconscious to use. The goal of psychoanalysis, for Freud, is not to restore some natural and more original breath free from anxiety. Steps must be taken—there needs to be a sustained transformation of anxiety, and then an extension of the symptom into the world.

Let me put it this way: I am not looking for the liberation of breath but for the unique breath signature of the hysteric. This could be her manner of speaking through drawn-out hesitations, almost a stutter, only to erupt in song. Her emboldened research into the failures of Western medicine, the invention of new modalities of care. A dance choreography that seems to show us the claustrophobic body of nine million human inhabitants on earth. The fever pitch of a writing that is faster than the internet but whose speed is decisively human and not machinic. Or a mode of gathering bodies in protest that can hold at the point of extreme pressure from the outside. This is as much a new pattern for breathing as it is a new language for life.

WHO HASN'T WOKEN up with a gasp, searching for breath, heart pounding against your chest while coming out of a nightmare? I've had patients who tranquilize themselves to sleep to avoid these ferocious visitations. That means also forgoing the other nighttime possibility, the wonderous, magical dream, a Faustian tradeoff they have decided is worth the sacrifice.

I'm not sure I agree. The maverick British psychoanalyst Wilfred Bion thought not dreaming would do something to one's capacity for imagination. I don't like dropping into and out of blank sleep. I'd rather be hounded and pursued, disgusted and dismembered, with death looming like a carrot on a stick. Nightmare in Hungarian is *boszorkany-nyomas*, "witches-pressure," linking these difficult dreams to the feeling of constriction in one's chest, the fright that takes away our breath. How alive your body can feel afterwards!

Nightmares are a reminder of what it means to be living and breathing. What a spectacle of life's derangement takes place in nightmares, such an array of near-fatal errors, or the incessant nature of fears around the ubiquity of human aggression. So much violence. Where does all this rage and terror come from? I think many psychoanalysts believe, little pyramid makers that they are, that rage feels better than fear, and fear feels better than the recognition of loss. Each of these is accompanied by its own experience of breathlessness.

Nightmares are close to the conscious experience of anxiety, though intensified in manifold ways—heightened visually, accompanied by bizarre sensations in the body, or imbued with extreme feelings of disorientation. I've had plenty of nightmares even if I don't have all that much of a conscious sense of anxiety.

On the surface, nightmares seem to run counter to Freud's basic dictum that the dream is a wish. But somewhere in the nightmare is the wish, even if it's just the wish for mastery of something unpleasant. Locating the wish, like echolocating our deepest desires, would put a stop to the nightmare. I always liked this task, turning the nightmare inside out like a glove to see its soft, colorful inside. Finding the nightmare's nestled secret, turning it out toward the world. This transformation of the nightmare into the wishful dream is similar to the transformation of anxiety into hysteria, or anxiety into desire.

Conversely, Freud notes that for many people what ought to bring great pleasure can often bring anxiety, confusion, and a whole complex of negativity. Freud discusses this in one of his last inquiries, an open letter he wrote to his beloved friend Romain Rolland (the same friend who attested to the feeling of oceanic oneness) at the age of eighty-one, called "A Disturbance of Memory on the Acropolis." Already it sounds like a dream.

At the start of the letter, he says he has often returned to a memory that he wants to finally analyze. Freud and his brother used to travel together for holiday when they were younger. On one occasion, thirty years earlier, the pair had planned to visit the island of Corfu but were told by a friend that it was not a good idea and instead they should consider going to Athens. Before the travel offices had even opened, they immediately became depressed, arguing with one another that it would be full of difficulty, if not impossible, staying in this irresolute state even after they successfully booked passage. The story culminates in Freud, standing on the Acropolis the next morning, overlooking Athens and the glittering sea and exclaiming to himself: "So it really does exist!?"

This thought felt disturbing. Freud was in his fifties at the time. Did he really doubt the existence of Greek antiquity? He'd spent his life studying ancient Greek culture. How could it be possible that there was a deep kernel of doubt the entire time? Freud suspects that the thought was about something else. Settling down to work, Freud searches for the link between the depressed state, the fog of pessimism that descended, and the feeling of derealization. Something was being expressed, something like the feeling: there is *no way* we will ever get to see this place we dreamed of as children. One might think of this as the opposite feeling to oceanic oneness—such was Freud's sardonic joke in his letter to his friend.

It's a paradoxical moment because often, Freud notes, the mind wants to avoid something unpleasant. But here, the mind is defending against something that would seem pleasurable, protecting itself from a great success—to have made it so far, to have seen with your own eyes a place of one's dreams. Freud recalls that growing up under great poverty as a child, he longed to travel. This longing is often the expression of a wish to escape from the pressures and unhappiness of childhood, like the "force which drives so many adolescent children to run away from home." The wish, surmises Freud, holds within it an aggressive criticism of one's family that induces guilt and makes certain pleasures forbidden.

Aha! So, the disturbance of mind and memory was the result of guilt. His father, he notes finally, had to go into business early and did not receive a secondary education. He wouldn't have enjoyed visiting the Acropolis. Freud surmises that his pessimism, the feeling of limitation and unreality, was nothing but *filial piety.*" He wonders if the pessimism that overtakes so many of us has this guilt and rage as its source, a guilt over aggressive criticisms we have of our elders who at one time we admired like gods. We love our parents, so we will not enjoy what will be at a cost to them.

Is piety, an infinite demand for dutiful conduct and unthinking reverence, what is being asked of us in our nightmares, those dreams filled with gods and monsters who appear and command us to surrender our lives to them? Do what I ask, and you can keep your life. In this sense the nightmare, at least on the surface, seems to run counter to another of the most important concepts in psychoanalysis—ambivalence, the fact that we love the people we hate and hate those we love.

The fact that elders who command our love also stir more violent feelings is one of the most important sources of ambivalence. The nightmare, with its bogeymen, makes the task of seeing our complex and ambivalent identifications difficult. Even as nightmares might, at a deeper level, invite us to renegotiate these splits, on the surface they tend to make them feel more substantial: "I'm not a monster like *him*."

Not long ago I had a nightmare of a mass shooter who shot my left breast. Left is where the heart is. In a bombed-out parking garage, I crawled towards a pool of bright green water, reminiscent of the Caribbean Sea, and woke up before reaching it.

The day before this nightmare, I had listened to a podcast on the awful suicide missions of school shooters who kill before finally killing themselves. My partner said these school shooters were cold-blooded killers; something about this seemed to annoy me, which was surprising since I myself often felt this way, especially when listening to their pre-planned machinations. We ended up fighting about it; I suppose I thought he was simplifying things, turning it into a question of good and evil. In the dream, this conversation became transformed as I cast myself as the victim of the shooting, as if to uphold his position rather than undermine it.

The following day, as I thought about the dream, the lyrics to a song by LCD Soundsystem came to mind: "Dashing the hopes / smashing the pride / the morning's got you on the ropes / and love is a murderer / but if she calls you tonight, everything is alright . . . *I can change, I can change, I can change.*" The song begins with the chorus,

"Never change, never change, never change," flipping the demands between romantic partners as they circle love and hate. I remembered that many school shooters often attempt to contact a loved one in the days leading up to that fateful day, as if to try one last time to find a benign presence that can right the situation. What does it take to wrestle with the ambivalence and aggression we feel towards others, especially loved others?

Seeing the ocean like that was not unusual. For years, clear pools of water had appeared in the middle of anxious nightmares, bringing relief. Connecting this with childhood memories of playing at the seaside, my analyst suggested I was looking for an unobstructed pleasure, one that felt clear, maybe even clean. Like a good clean fight—if there is such a thing. She seemed to indicate with her tone that I should consider whether this was really something I could ultimately find.

"I'm going to die trying!" I retorted. Then I finally heard it. The suicidal pursuit. The wish for a clear conscience—the very thing I had condemned in my partner. Turning the nightmare inside out, I came to see the wish embodied in the nightmare: teasing and taunting with that glimmering pool of water. The water was expressing the desire to be unburdened of all my own feelings of violence. Lacan invented a word for this—*hainamoration*: a portmanteau of *enamouré* ("in love with") and *haine* ("hatred").

Garage means "shelter," coming from Old French *garir*: "to rescue, take care of, protect." What else am I sheltering than my own loving-hate? The psychoanalyst Christopher Bollas points out that innocence is violence, aimed to keep us away from troubling recognitions, especially regarding our own injurious self-protection. The violent innocence provokes the other into a useless frame of mind, a feeling of extreme psychic isolation. Children will often spare the adult whatever troubling recognition by joining them, rescuing them, and abandoning their own mind. The solution is a shell, "a dehydrated" internal world. What a perfect dream image then: an empty garage and the search for a cool pool of water. So here is the wish—to protect me and those I love from myself.

WILFRED BION IS the only psychoanalyst I know to emphasize the psychic life of the fetus. "The foetus has no choice but to get born and is forced out into harsh gaseous fluid instead of a nice watery fluid," he claims in an exchange in New York in 1977. Perhaps detecting the remnants of a certain oceanic fantasy, Bion's interviewer asks him if water is always better than gas. "Not always," he replies. "The foetus is clever enough to take a little of the watery fluid in the nasal channels. The result is that it is still able to breathe and smell. Smell travels very well in a watery fluid; it is a long-distance receptor. Fish can smell decaying matter from a distance of many miles."

Bion goes on to point toward the possibilities for communication created by our having been born, saying that psychoanalysts "have no choice except to communicate our interpretation by virtue of the gas—air—which we use or abuse for phonation."

For Bion, intrauterine existence is already a site of impingement and defense, creating thoughts that never quite make it to consciousness but exist in us for the entirety of our breathing life. We cannot remember the time we spent in utero, in the same way we can't even really remember being infants or young children. But we were there, and it has left its mark on us. Bion called the fetus "very intelligent," and indeed we now know that fetuses not only smell, taste, and see, but also shield themselves from bright light, recognize the voices of their significant others, play with their own umbilical cords—enjoying the sensations and rhythms of them as different to those of their own bodies—and suck their thumbs. They even seem to prefer some music over others.

Think of how this opposes the image we have of the fetus as a naïve being living an existence of bliss, not having to breathe or eat, enclosed in a warm watery surround—dreaming and connected. Against nostalgic fantasies, Bion sees the womb as a place where we were not easily at home. The fetus's intelligence develops because it is already hounded by sensations so easily transmitted through water. None of this can be controlled by the fetus, except via beginning to develop intellectually: "I see no reason that the fetus isn't born with a personality," he exclaims.

Following his mentor, the child psychoanalyst Melanie Klein, Bion says the fetus in the womb has already started developing the capacity for "projective identification." Projective identification isn't what we tend to think if we've been introduced to the term in pop psychology. Most imagine it as a process whereby I attribute my thoughts or feelings to another person, and they end up acting as if they did, in fact, belong to them. However, the idea is much stranger than this typical porousness in human relationships.

Bion points out that projection is a fantasy that one can get rid of parts of oneself by hiding them in the other person. Klein calls the anxiety that leads to this defense "psychotic" because this creates a real fragmentation of identity. You don't know what belongs to you as reality is more and more distorted. The body or world begins to feel like what Bion called a "bizarre object." This is like a synesthesia of our sensorial and cognitive capacities, like a wall that feels like it is breathing heavily and watching you, or skin that feels like the surveillance system of a porous wall.

Splitting was one of the last topics Freud wrote about before his death. In a 1938 text he zeroes in on the fact that reality could be split in half, so that people can really act as if they believed two contradictory ideas at the same time. Bion sees the mechanism of splitting as following an evacuation into the outside, dumping something unwanted into an exterior, toxifying it. He likens this to the fetus polluting its own waters (that transmit all these impinging sensations) with meconium. You then must look through, feel through, breathe through, your own shit.

This follows a fascinating and offhand remark by Freud in his *Introductory Lectures* where he says he borrowed from the "naïve popular mind" the link between anxiety and birth. When asking midwives what it means if a child is born with black water (from meconium), they answered that the child had clearly been frightened. While the doctors scoffed, Freud said he sat in silent agreement.

Already in the womb we began destroying ourselves and Mother Earth out of terror! Being scared shitless takes on a more literal tonality. The fetus expelled its insides prematurely and mortally endangers itself. This is a significant hurdle for psychoanalysis because it is difficult to rush in and save someone from themselves. The patient must want to live. There must be at least a minimal integration between the half that knows it is poisoning its environment and the other half that denies this truth.

Bion compares splitting to the diaphragm, which divides the human being in half. The diaphragm at the bottom of the lungs, he says, is where all the arguments rage about the thinking done by the body and that done "at the cranial" end. The diaphragm, Bion suggests (one can never tell how literal he is being), is what links these two halves of the self: it moves up and down, is permeable, and is a source of circulation between all parts of the body.

The diaphragm shows us that we must try to see everything from both sides, from the bodily end and from the cranial end. From the top and from the bottom. When someone speaks of an inhibition or anxiety, what it means from the position of yes, and what it means from the position that says no. As Joni Mitchell famously sang, "I've looked at clouds from both sides now, from up and down and still somehow, its cloud illusions I recall, I really don't know clouds at all."

How easily with any given patient can you "change your position, your vertex, so that you can almost see both sides?" Think of my fight with my partner about school shooters. This wasn't an intellectual argument about the school shooters' culpability and responsibility more generally, in which case we would have to think more seriously about gun laws in America and our horrendous mental health care system. Rather, with someone I love desperately, I was in conflict around a split that the image of violence provoked where condemnation of some was always a declaration of innocence. I wanted us to be on the same side and found myself stranded on another shore. I wanted to condemn the aggression of condemnation. This is a moment of Bionian madness.

When we split, it's as if the diaphragm isn't working. Maybe the diaphragm is barely moving, just holding the division rigid and taut. Or maybe it's pumping so fast you can't get a sense of what belongs where. "Psycho-somatic or soma-psychotic—it's the same thing: the same impressive trauma, looked at from different sides." I just love the idea of the soma-psychotic! Analyzing asthmatic patients, Bion exclaims, will lead to all kinds of trouble. The soma-psychotic side of the psychosomatic means there is an inaccessible catastrophe that has been got rid of "*at source.*" While neurotic headaches, stomach aches, racing hearts, and difficulties breathing can seem like everyday psychopathology, Bion is pointing to a deeper side of these phenomena that attests to feelings of catastrophe and damage that we barely let ourselves know about.

As if to put this strange assemblage of thoughts together, Bion finally links the idea of the intelligent fetus and the diaphragm. Freud and Otto Rank, he notes, both spoke of the "impressive caesura" that is the trauma of birth. For Rank, life requires at least two experiences of rebirth, one from the mother's body, and then again into an independent identity through some heroic act of will that surmounts anxiety. Bion, on the other hand, suggests there is a continuity between the fetus and the person who will eventually come to use and abuse the air; no need to be reborn. We could become a little less "born" in fact. Less sharply individuated. Work with the caesuras that make it difficult to know parts of ourselves.

To insist on the continuity with fetal existence means to step back from this mode of projection. Intrauterine life is not some mystical, seductive, or terrifying foreign land. We have to think our way back behind the screen. Bion's idea of a capacity for reverie—for picking up subtle thoughts on the edge of consciousness, for daydreaming—takes shape in thoughts on fetal existence.

The lecture in New York ends with Bion reminding his audience not to venerate thinking: if the fetus is already thinking, then our idea of thinking isn't what we think it is. Psychoanalysts are witness to the impotence of thinking. The thinking that is breathing would be a better place to go, bodily, diaphragmatic, closer to this caesura between us and whatever catastrophes linger in our psyches.

Bion was Samuel Beckett's psychoanalyst. In thinking of Bion, I can't help but think of Beckett's minute-long play *Breath*. The lights turn on and dim along with a single exhalation and cry, a primordial scream (of birth?) on a stage covered in litter. Or *Not I*, a monologue where you only see the actress' mouth, which unleashes a torrent of words, audible breathing used to keep the mouth speaking what are never entire thoughts but passing fragments that cannot be integrated into an "I": " . . . out . . . into this world . . . tiny little thing . . . before its time . . . in a godfor . . . what? . . . girl? . . . yes . . . tiny little girl . . . into this . . . out into this . . . before her time . . . godforsaken hole called . . . called . . ."—and I'll jump—"thin air."

I like Bion's speculation (as he would put it) on the origins of life and breath. He never quite converts this to a body technique, but he keeps the body there, part of our psychotic-somatic core. Bion gives the trauma of birth a different valence, free from the fantasy of a blissful state before life or after it. The division goes all the way back. An ongoing constitutive division. One that requires constant work. Diaphragmatic pumping. You must get curious about yourself, about what you don't know and don't want to know. Start attending to this and force your awareness to circulate in your thoughts.

ENCOUNTERING THE ANXIETY of my patients always felt slightly alien. It sometimes inspired disbelief—as if I thought that they couldn't really be so fearful, so limited by their worries. I've learned to see this as my own problem, as if the anxiety that I can't recognize in myself creates a blind spot on the outside. Finally becoming anxious in analysis ultimately solved this empathic failure. My first supervisor called me counterphobic.

It wouldn't take all that much time for the counter to resolve into its true face—anxiety! I started having trouble in social situations, tormented by the feeling of being judged. Looking back, certain brazen decisions in my life to change course seemed laced with rationalizations— was I merely running away? I couldn't look at drafts of papers for fear of seeing grammatical mistakes. What on earth was going on?

I was furious. This is where therapy gets me? This paralysis? Here, we might recall Bion: "In every consulting-room, there ought to be two rather frightened people: the

patient and the psychoanalyst. If they are not both frightened, one wonders why they are bothering to find out
what everyone knows." As consolation, I charged myself
with the task of finding real courage rather than continu-
ing to rely on hysterical acting out.

Another macho dare—I'm not afraid to look my anxiety in the eye!—but it is good to get to know this problematic puppet master behind the scenes, to unearth even
the hysteric's anxiety. There are new possibilities of love
and care that haven't been accessed, needing to face what
is terrifying and unknown.

Before analysis, I'd had a few fleeting encounters with
anxiety, like the horrendous fear I'd get after smoking
marijuana. It being substance-induced anxiety, however,
made it feel like it wasn't really *mine*. Marijuana brought
on pure existential confusion: What am I doing, why am I
here, what's going to happen, what do people think of me,
what does anything mean, what have I even done with
my life? Time just felt impossibly long. Too long to live
with these impossible questions.

Disorientation leaves you so naked. It's as if I'd lost the
compass I'd been using to orient myself. The heart palpitations then kick in. The difficulty exhaling follows the
increased autonomic arousal so characteristic of anxiety
and panic. Deep breathing can do wonders at this place,
taking you back inside and giving you a sense of control.
Refocusing on the rhythm of the breath is important to
put rhythm back, like a container, whose force of separation and beginning differentiation works on an anxiety
without boundaries or edges.

I'd had one other disorienting moment of anxiety when I took up surfing for the first time. I had never been thrown around by the ocean like that. Tumbling underwater, not knowing which way was up, while the board went flying who knows where. In those seconds, you wondered when all this was going to stop and whether your breath would give out before you found the surface.

What's funny is that I don't remember feeling consciously anxious. I just had the feeling that I didn't like surfing, didn't like that I felt so dependent on my (mostly male) instructors. Or that I didn't like this kind of ocean. I'd grown up with green see-through Caribbean waters: going underwater was always associated with feelings of calm and wonder. A lightness, where breathing buoyed you or helped you descend.

Surfing was something else: the sea as pure power—unpredictable, chaotic, and violent. I wasn't in charge of my breath, the sea was. And I did not have any means to breathe underwater should it decide to hold me there. Eventually, in some defiant gesture of trying to surf without an instructor, I paddled around for an entire week without really trying to catch any waves. So, I was never forced underwater into a state of disorientation. I claimed that I didn't like surfing in Long Island. The waves were small and sloppy and short. They were. The water was crowded with too many surfers. It was. I deferred surfing again until I returned to Costa Rica. (I'm still waiting to get back there.)

Putting it like this, seeing it from the outside, it is easy to say there is palpable anxiety. From the inside, it never felt like that. I simply didn't like surfing and wanted better conditions. I even claimed I wanted bigger, stronger waves.

This is really the essence of the difference in psychoanalysis between what we call anxiety hysteria and conversion hysteria (or *hysteria* hysteria). Anxiety hysteria is a failure of the transformation of anxiety into a symbolic world governed by desire's network of wishes and memories. Desire is a terrain that can be surfed thanks to the semantic overhaul of anxiety. Our symbolic system is replete with attachments to people, places, memories, things. In this case, my childhood ocean. But this wishful constellation hides something darker beneath its surface, providing cover for a breathless childhood anxiety.

This overhaul is about preventing the bodily feeling of anxiety, but it is also so much more than that. The ways that fears and delight function through a symbolic system are the missing guardrails that point us towards some things and away from others. The word *conversion* speaks to this transformation—the conversion of anxiety away from its amalgamation of mental dread and hyperactivation of the sympathetic nervous system into a symbolic symptom that depends on repression. This is an act of personalization: you have *your* symptom, not this diffuse symptom of mind and body gone haywire.

Freud called hysteria a capacity, a talent for getting rid of anxiety, transforming it into the stuff of desire. But what is repressed? What about those lingering, hidden inhibitions and anxieties? This is the real stuff for psychoanalysis. Circumscribing anxiety creates a symptom. Symptoms can be analyzed. What are you still afraid of or too excited by? What in this can you truly desire? What do you misunderstand about yourself here?

I had been left alone as a child and I had all the armoring of a defiant self-reliance. There were a thousand and one excavations of this terrified, lonely, and depressed child that I had to unearth. The memories I had to unlock still astound me, along with all those I simply cannot. Connected to these rather traumatic memories was a resilience and way of leaning into desire that I inherited from my family despite everything. The traces of this desire became more and more apparent the deeper I delved into my anxiety and helplessness.

Over time a wish and inhibition came to the surface. I wanted to have another child, a daughter in particular, and for some reason I couldn't find the will to let myself. Even worse, what I really wanted was to have a second child with a partner to raise her with. One might utter the dreaded word *family*. I don't really want to, even now. This was simply too much dependency. Too much togetherness—yech!

Something obscure rose up like a thick fog around me. I just couldn't see my way through. A boundary seemed set in stone and any attempt to approach it induced life-death panic. A line that spoke to a stillbirth before my birth, a vague decree that having children was expensive and imprudent, some image of what it meant to be a "good" girl (i.e., not sexual), a criticism of my family where I was the only child, a feeling that I would jeopardize *my* career. How could I—after all these years of analysis and working as a psychoanalyst—be held back by all this old stuff?

Was all the analysis I had done prior to this just a façade? The honeymoon period, as we like to call it, before the real struggle kicks in? During training, my teacher asked the class, "How do you know you are in middle-phase psychoanalysis?" "SNAFU," Dr. Sax said with delight: Situation Normal, All Fucked Up. In class we would read patient intake reports and then guess whether the patient's psychoanalysis went well or not based on the termination reports (which Dr. Sax wouldn't show us). We were, of course, always wrong. What it takes to really get to so-called later phase analysis will never be clear from someone's initial presentation. Dr. Sax was trying to disabuse us of any pretentions of knowing what makes for a good patient, also demonstrating how difficult and long psychoanalysis is.

I always remember this term—SNAFU. It's believed to have emerged from the Marine Core in World War II. My father was an air force pilot in the marines before he became a commercial pilot, managing to serve in the space between the Korean and Vietnam Wars. Even though they were worlds apart, Dr. Sax seemed to me like my father's kind of guy. Tough guys. Traditionalists. That type was everywhere at the psychoanalytic institute I ended up in (of course). There I was, having taken a completely different path from him yet surrounded by guys just like my father: situation normal, all fucked up. I was anxious and inhibited and filled with conflict. A shallow breather in life! What would it take to confront this desire without cover?

I felt for the first time a requirement around desire: there are some things that come without any justification or sanction possible. No matter how hard I tried to find it, I couldn't. Believe me, I tried. I wasted years trying to find it. For someone seemingly so ruled by desire—almost automatically—here I was, unable to just want what I wanted. Never had I found myself so tormented, framing a long period of struggle in the latter half of my second analysis, wrestling with the most obscure and anxiety-inducing conflict (what *piety!*) in the pursuit of a family. "Who do you think you are?" the voice in my head uttered. To which I could only hold my breath.

The diffuse anxieties that entered my life at a certain moment of psychoanalytic work crystallized into this one anxiety and inhibition about motherhood and family. The prohibitions against this desire, the images that filed in and made it impossible to pursue with any straightforward vigor, showed the force of condemnation and allegiance that can trouble a mind. Close to the end of my psychoanalysis, this was certainly a powerful representation of parental authority (and no doubt deep counter-criticisms of that authority) that needed to be confronted. What would it take?

In *Otherwise than Being*, philosopher Emmanuel Levinas asks, "Is man not the living being capable of the longest breath in inspiration, without a stopping point, and in expiration without return?" This was my longest breath. I didn't know what I would gain from it, what return to myself I would find. By luck, I returned with her in my arms. Having a daughter made me part of a lineage of women. This is not uncomplicated. But it gives me a new birthright by shifting ever so slightly my name, the son of James. As her father likes to call her sometimes—Jamiedottir.

PULMONAUTS, AS JAMES Nestor calls them in *Breath: The New Science of a Lost Art*, have come and gone for the last two hundred years—those who have been hailed for discovering, or rediscovering, this basic breathing cure for modern illness. The possibilities of breath have repeatedly been forgotten, thus having to be reinvented and pushed by new figures. Nestor points out that almost all these figures were autodidacts, each desperately searching for a missing truth that seemed to be right there in front of them, hiding in plain sight—like air.

Perhaps the most common piece of pulmonautical wisdom concerns the benefits of breathing through the nose; these recommendations go as far back as an ancient Egyptian medical text from 1500 BCE declaring that the nose should be the gateway to the heart and lungs, not the mouth. This wisdom reappears throughout many religious or spiritual practices but doesn't reappear as a concern for modern Western populations until the nineteenth century, long after our descent into mouth breathing.

These practitioners-cum-anthropologists-cum-gurus were themselves always at risk of getting wildly lost in the battle for recognition of something that didn't seem to want it. Reading about the fallout of the rediscovery of the power of breath started to scare me. The stories amount to a tragic, heartbreaking feeling of what humans must learn and relearn about breathing.

Even James Nestor's Instagram was upsetting to me. A wonderful writer and journalist, he had become a one-line proselytizer of nose breathing from a book that was fading into the distance. Maybe it's just Instagram, though breath does seem to have somehow become a commodity sold as a means of curing twenty-first century alienation, with a long and forgotten history of encounter and self-cure. Was this going to be his final cross to bear?

Katharina Schroth, a teenager in Dresden in the early 1900s, had severe scoliosis. She devised a way of forcing air into one lung at a time and then stretching her body across a table in back-bracing positions to expand her rib-cage. After five years of doing these exercises, she cured what is considered an incurable disease that would have left her bedridden. She literally straightened her own spine through breath. Her knowledge follows an ancient Chinese adage from 700 BC that what follows perfect breath is perfect form.

Carl Stough, a famous choir director, did breathing and singing exercises with veterans suffering from emphysema in a random bid for help in 1958 by a hospital in New Jersey beset with chronic patients. Seeing that their problem was one of exhalation, he mitigated the worst of their disease through radical techniques that taught them how to release stale air by pumping with their diaphragms. Patients were walking around within weeks of his treatment.

Stough was also responsible for twelve medals and five world records for running in the infamous 1968 Olympics in Mexico City (where several athletes who were also Black Panthers raised their fists upon acceptance of their medals). The sprinter Lee Evans learned the art of exhalation from Stough. Regardless of these accomplishments, his work never caught on. "He made his map over half a century of constant work," writes Nestor, "and when he died the map was lost." Emphysema is still listed as an incurable disease.

This list of miracle cures repressed, mocked, forgotten throughout the centuries goes on and on. Reading about this "lost art" of breathing, I eventually started to worry about myself and my own book, about my profession with its claims to cure and its supposed airiness. The nose appears in a primal scene in the origin of psychoanalysis. Freud tried to cure his neurosis via nasal surgery with Dr. Wilhelm Fliess, Freud's early mentor, who also acted as a pseudo-analyst while Freud struggled with his new theory of the unconscious. Freud wrote to Fliess about his discoveries, his dreams, and his work with patients.

We do not have Fliess's side of the correspondence; Freud presumably destroyed it. Freud wanted *his* letters to be burned, too, sending the wealthy princess and psychoanalyst Marie Bonaparte to buy them back off Fliess's widow. Once she had acquired them, however, Bonaparte refused, since she recognized how precious they were, revealing how in the early days of psychoanalysis, Freud struggled with his ideas by wrestling with his own neurosis in a powerful transference to an older male authority figure.

Fliess was monomaniacal, much more so than Freud. Fleiss was obsessed with supposed nasal menstruation as evidence of bisexuality, his pet topic; he asked Freud to record in detail the days his nose filled with mucous and eventually suggested nose surgery for his depression. Under Fliess's watchful eye, Freud eventually began to elaborate his own theories, paying special attention to the nose. As humans evolved to stand upright, the nose became less important, and therefore less sensitive, as a means of interacting with the world. Freud thought that the drives connected to smell were inhibited or repressed. We no longer routinely smell each other's genitals, though some neurotics, he found, hold on to their acute sense of smell.

In *Your Inner Fish: A Journey into the 3.5-Billion-Year History of the Human Body*, the scientist Neil Shubin writes:

> Humans devote about 3 percent of our genome to odor genes, just like every other mammal. When geneticists looked at the structure of human genes in more detail, they found a big surprise: fully three hundred of these thousand genes are rendered completely functionless by mutations that have altered their structure beyond repair.

By studying the DNA of different species, we can see that smell shifted with the transition from water to air, and then again with the development of seeing in color.

As Shubin writes, "We humans are part of a lineage that has traded smell for sight."

However weird their exchange, Freud and Fliess were perhaps onto something. Freud later found that overdependence on the visual appears very strongly in neurosis, with its extreme impulses to look, peep, and surveil—or, on the flip side, to exhibit, display, show, perform. This reliance on the visual is especially acute in the psychoses where one can feel observed in a persecutory way. If one wants to breathe deeply, it is always best to close the eyes or at least soften one's gaze: to suspend the tyranny of the visual. Meditational postures often show the eyes half-closed or turned downwards. Lacan notes that Buddha's eyelids are half-closed to avoid the trap of seeing. In the practice of yoga, one is told to soften the gaze in order to maintain balance, to learn not to use your eyes as the only way to stabilize your body and breathing.

When it comes to breath, the promises are infinite. Relearning to breathe through your nose, the foundation for a whole range of breathing practices, promises a panacea that cures snoring, sleep apnea, chronic congestion, asthma, anxiety, emphysema, autoimmune disease, erectile dysfunction, misshapen faces, narrow jaws, and crooked teeth, while also making for athletic superiority and, finally, spiritual enlightenment. Controlling the left-right breathing through one's nostrils can reintegrate the split halves of the brain and balance the mind and body. Occasionally, certain breathing techniques are said to cure schizophrenia. Generally, changing breathing is hacking the body, altering the nervous system, controlling the immune response, and restoring health.

Not all of this is nonsense. The tales of what swami priests can achieve are awe-inspiring. They are indeed able to raise their body temperatures, slow their hearts to near-coma conditions, turn off their amygdalae, hack the parasympathetic nervous system, and breathe slowly—really, really, slowly. And sometimes fast. Very fast. Hyperventilation, which increases carbon dioxide saturation, is also part of this curative tale.

The wisdom of pulmonauts always culminates in attempts at overcoming whatever about modern civilization has left us pitiable mouth breathers, slovenly in our work-weary disenchantment—pure mind, no body. Searching for God. Dreaming of the apocalypse. "The way a man sings to himself in moments of dazed, thoughtless excitement, without even knowing—and he uses what tatters of breath he has left," says the narrator of *The Magic Mountain* of its patient Hans Castorp on the eve of World War I.

Who, staring at the skulls of our ancestors, with their perfect teeth, large mouths, unblocked airways, who thus breathed easy, wouldn't begin imaging a utopia distant from the chaos of contemporary life? Why have we become sick? Why would we evolve to not be able to breathe?

Scientists are trying to find out. One idea is that we are in a process of dysevolution. But this declaration is speculative. What do we imagine is the telos of evolution to surmise that we are in a process of devolving rather than evolving? Maybe this not-being-able-to-breathe like *they* did is part of evolution? Anthropologist Robert Corruccini called diabetes, heart disease, and stroke the "diseases of civilization." We could add asthma and other chronic respiratory conditions to the list, not to mention the countless diseases caused or exacerbated by air pollution.

Certainly, we are more sedentary than our hunter-gatherer ancestors. Sitting hunched over a computer screen all day is causing unusual forms of strain—affecting our skeletal integrity, nervous system, eyes. Our ancestors also apparently spent a lot of their time chewing unprocessed food, making for their larger mouths, jaws, and perfect teeth. From examining skulls from around the world, scientists estimate that around three hundred years ago a transition in diet caused a near viral disintegration of the human palate.

On the other hand, we are also more mobile. The nose and skin adaptations to the environment don't make a lot of sense in a world where humans rarely live, generation after generation, in the same location. People with narrow noses and whiter skin for colder climates can park themselves in equatorial heat without the right equipment, so to say. Evolution hasn't caught up with this new human mobility. But that's where the technology enters of orthodontics, ear-nose-and-throat medicine, gyms, air conditioning, and sunscreen.

I'm not saying that this life is better than that life, or vice versa. I was taught as a psychoanalyst to try not to judge any manifestation of culture as better than any other. Our judgments most often support our illusions regarding happiness, quipped Freud. What I want to question is the image of a more natural, well-functioning life attuned to nature, which bypasses the part of evolution where humans banded together, moved around more, and began inventing forms of technology.

Ferenczi noted that human genitals were a kind of technology, or proto-technology, for the transmission of information in the form of DNA and a new protective container for embryos. They developed in response to leaving behind life in the ocean. Technologies, like genitals, are unavoidable and bring discontent. In the case of sexual differentiation, this development brought in physical domination by one half of the species, and civilization meant not merely sexual war but war between neighbors. We are thus always in the middle of some process of evolution whose end we will never see or know.

Human civilization, at least as Freud came to think of it in *Civilization and Its Discontents*, is a disjunction in our relationship to nature—impossible to go back on. Certainly, with all the inequity and destruction we have wrought (despite all our technological marvels), we look back with rose-colored glasses, as if there were some earlier moment of bliss that we are missing out on. Do we really think things were more equitable at some earlier time? Shouldn't utopia be something to seek in the future rather than the past? For psychoanalysis, getting closer to the reality at hand is the most important and most difficult task.

A fact about our breathing problems that gets very little attention is the evolution of language. "Talking comes at a steep price," writes the scientist Neil Shubin. Or in the poet Mark Strand's words, "breath is a mirror clouded by words." Our expanded language capacities caused the larynx to lower—which is mostly made up of "gill arch cartilage" linked to an evolution from the breathing apparatus of fish. The tongue sinks into the back of our mouths to more easily move it to produce sounds. The jaw juts forward and lips become smaller for easier manipulation. This evolution made humans more susceptible to sleep apnea because the more flexible muscles of the throat relax in sleep, which for some can amount to a collapse that stymies breath. The evolution also led to the increased risk of choking. As Ben Lerner pointed out in "Café Loup," we are the only species that can't breathe and swallow simultaneously. Is it such a wonder that we started eating more easily chewable food while we were more and more immersed in the circulation of language?

For Jacques Lacan, language is the most important idiosyncrasy of the human species—the fact that we are speaking-beings, as he called us. Language was a trauma that we suffered, a foreignness comparable to air that invades us from the outside. How do we assimilate language? As with breathing, we develop a personal rhythm in relation to this force that comes with its own history, forcing itself upon us. We are born into language that will outlive us—another narcissistic blow.

Lacan wondered if psychoanalysis had neglected a respiratory drive. He likened the trauma of language to the trauma of sexuality; we invariably use one to talk about the other, and breathing is also involved in sexual experience. Lacan noted that spasms of the diaphragm, larynx, and esophageal sphincter are involved in orgasm. This follows the evolutionary aberration going back, yet again, to the transition from sea to land. The part of the brainstem that controls respiration is close to the gills of the fish, but now that the diaphragm controls breathing, there is a long way to go to the brain.

This whole process is easily interfered with and prone to spasms. Amphibians breathe by closing their glottis, a flap in their breathing tube, through a kind of hiccup, a rhythmic pattern of stimulation which is then stopped by breathing deeply. This makes its appearance in mammals who sometimes hiccup without the benefit of being able to breathe underwater. But this rhythmic patterning appears again in the breathing and spasms of sexuality.

It is not clear where language and human forms of sexuality are taking us, nor where they are taking our bodies. As Michel Serres says, "the body is not only for itself, it carries in itself those that follow as well." If there is a respiratory drive, it is exercised most of all in psychoanalysis: say anything, say everything. But this respiratory drive must be thought of as being like the other drives, caught up in the vicissitudes of the evolution of the body and human sexuality—such that psychoanalysis, that use and abuse of air, is a cure without promise.

THE ABILITY TO hold one's breath underwater cannot be related to physiology alone, though some people do have larger lungs or larger spleens (crucial for flooding the body with oxygen-rich blood cells). Still, mostly it is a learned skill. Instead of triggering involuntary breathing, the free diver slows their heart rate and metabolism, redirecting blood to vital organs and bolstering the lungs against underwater pressure. The current record for free diving with fins, but without technological propulsion, is over four hundred feet, roughly the length of a forty-story building, up and back.

To accomplish such a feat, one must push through a limit that is always described as psychological. You learn not to give in to panic long enough so that the blood can desaturate itself of oxygen and begin a new process. Free divers speak of the "free fall"—the point when they reach a depth where the pressure compacts their lungs and begins to suck them down—a deep meditative state that defies the laws of the breath. Some speak of the feeling of quiet and calm that overcomes them—a serenity like no other. They break world records. They also often die.

After entering the period of free fall, the diver must still exert enormous effort to swim to the surface again, swimming *against* underwater pressure this time. What a task to set up for oneself—going under, way under, and then having to make your way up and out. Most emergencies take place in the last thirty feet before surfacing, when the diver can black out or suffer the effects of ruptures in the tissue of the lungs. In a sense, it is a real mastery of anxiety. But surely something in this mastery is death-driven, something that pushes these mystical athletes to descend to unheard-of depths and to withstand deathly pressures.

A major breakthrough in the understanding of such processes came in the late 1860s, when a physiologist named Josef Breuer started working on the question of respiration, eventually publishing a coauthored paper with Ewald Hering titled "On the Self-Steering of the Respiration."

Breuer and Hering left a mark on medical history; with the mechanism that became known as the Hering-Breuer reflex, they rediscovered the importance of the diaphragm, the ways we steer our own breathing through deeper or shallower breathing, changing the way scientists viewed the relationship between breathing and the nervous system. The science at the time held that oxygen and carbon dioxide levels in the blood were the only ways that the vagus nerve could be triggered, and that this in turn triggered breathing in a simple feedback loop. If breathing was self-steered—we moved the diaphragm and the diaphragm triggered the vagus nerve irrespective of blood chemical levels—then this opened a much more radical gap in the circuit. Placing self-steering in that gap meant the whole reflex could be changed, modified, hacked into, even perverted—thus opening up a much more complex picture of the relation between a person and their unconscious bodily processes. Of course, Eastern breath practices know this, but it took Western medicine some time to catch up to this ancient knowledge.

A few years after publishing this paper, Josef Breuer met a young neurologist named Sigmund Freud who was interested in the unconscious life of his neurotic patients. In 1895, the pair published *Studies on Hysteria*. In this groundbreaking work, Freud and Breuer described numerous cases of hysteria along with their new cure that went beyond hypnosis in the direction of unconscious thoughts and feelings that needed to be expressed, what would go on to be called psychoanalysis—or the "talking cure" or "chimney sweeping" as Anna O., one of Breuer's patients, called it. The psychoanalytic cure now feels like a remarkable extension of Breuer's work as a pulmonologist, even more than an offshoot of the new science of psychiatry.

Anna O. was the most important case in the *Studies*, with some of the most memorable symptoms, including nighttime states of self-hypnosis, a spasm of the glottis the moment she reproached herself for wanting to dance, a nervous cough called *tussis nervosa*, and finally a hysterical pregnancy that she attempted to deliver in Breuer's presence.

Rejecting the various forms of mumbo jumbo that were being used at the time—hypnosis, massage, water baths, rest—Breuer followed Anna O.'s inclination to talk. (That said, Breuer did use Seconal, a powerful sedative, leaving Anna with an addiction to it for which she had to be institutionalized. This was not mentioned in their case write-up.) It's for this reason that Freud credits Breuer with the discovery of psychoanalysis. Breuer's interest in self-steering mechanisms, their involvement with what was thought to be simply automatic, makes ripe the transition between breathing and hysteria.

We could speculate that Breuer's earlier research into breath work played a part in his and Freud's desire to make visible unconscious mechanisms, to correct what was perverted, and to find ways to intervene through the gap in the psycho-physical feedback loop. In *Studies in Hysteria*, cure no longer consists in hypnosis, drugs, or any purely physical therapy. Instead, cure is activated through a combination of talking and transference, using the influence a trusted doctor can have over a patient and speaking into the gap between the psychic and the physical.

But psychoanalysis became synonymous with Freud, not Breuer. After treating Anna O., Breuer went on his honeymoon with his wife, and he would never return to treating hysterics. Many psychoanalysts have pointed out how much Anna must have scared him. Her transference towards Breuer was intense, as was the sexuality implied in it. No doubt he struggled with his own interests in her—Anna O. had Breuer visiting her at all hours of the day, day after day. She began going into fugue states and would only come clear when he would visit and spend time talking with her. If it had been some time since his last visit, she would make it tougher on him to get the desired relief they were both now looking for.

The baby that she gave birth to in imagination was Breuer's, and she must have known somewhere that he was about to get married. "Dr. B's baby is coming," she screamed, writhing and panting on the floor. Talk about self-steering! The baby, this talking cure, would become Freud's. Breuer decided that he disagreed with Freud about his emphasis on sexuality (seen in letters) and backed away from the field of psychology to focus on his medical practice. He never publicly disagreed with Freud, but the two men were estranged.

Freud, it must be said, dived headfirst into the depths of transference and sexuality, refusing, in the face of such mysterious powers, to give in to anxiety. He mastered anxiety by confronting in himself what he was to hear from his patients, a swirling depth of confusion, yearning, pain, sexual passion, and aggressive complaint. As he famously said: "No one who, like me, conjures up the most evil of those half-tamed demons that inhabit the human breast, and seeks to wrestle with them, can expect to come through the struggle unscathed." To this image of the analyst as Virgil, descending into the depths of hell, we should remember our free diver in free fall who must find the strength to come back to the surface. Freud comes up for air, wanting to tell the world what it means to know that most of it is but the tip of an iceberg.

How CAN I write about breathing without discussing Eastern spiritual practices, yoga, and well-being? I don't really want to do it, perhaps because the way well-being practices parallel the expansion of psychoanalysis, especially in the United States, fascinates and revolts me.

I also have a problem with this fundamentally orientalist project, a project that in some way is part of my very being as an Asian woman whose father, of British and Swedish descent, escaped extreme poverty through military service and, as a fighter pilot for the marines stationed in Japan and Korea, developed a penchant for Asian women. These women must have been incredibly exotic to him, so very unlike his own mother, who he worshipped and who he saw ravaged by a bare life in the Midwest. He was furious when my mother cut her waist-length black hair.

Like psychoanalysis, yoga first arrived in the United States in early twentieth-century New York. There was seemingly no end of rich patrons in the Northeast looking for something new. An esoteric practice framed by images of sexuality, sacredness, and liberation, it was especially embraced by women. Fast-forward to the 1960s and '70s, and The Beach Boys and The Beatles were touring with their swami, while increased mobility saw a surge of Westerners making pilgrimages to India and Tibet.

Today, I can hardly escape the mantra to exhale, to give gratitude to my breath, to use my breath to ground or center myself, releasing my thoughts and finding the space for acceptance and non-judgment. The airy language of well-being is used to sell just about everything. Clothes that breathe. Products that increase compassion. Yoga mentality offers a generalized cure for the alienation from the body that defines contemporary life in the twenty-first century.

My problem with the industry is a problem with it as an industry—as a profit-driven assimilation of distinct, local histories. But I'm a guilty party. I've been practicing yoga for twenty years. I have myself made the voyage to India to learn. The anthropologist in me fights with the practitioner of yoga which fights with the psychoanalyst, and it is the latter that cannot fail to notice the neurosis so common among those who claim spiritual enlighten-ment: Why does all this have to come with such a mighty dose of self-righteousness? Mind you, I have the same complaint for the world of psychoanalysis. I think of Freud writing to James Putnam in America in 1915:

> The unworthiness of human beings, even of analysts, has always made a deep impression on me, but why should analyzed people be altogether better than others? Analysis makes for integration, but not necessarily for

goodness. I do not agree with Socrates or Putnam that all our faults arise from confusion and ignorance.

Modern yoga is very different to the ancient yoga tied to Hinduism, which is closer to meditation, breathwork, prayer, and reflection. In fact, sitting was the only asana. Ancient practice was aimed at cultivating prana, the invisible "breath of life." As Ken Rose and Yu Huan Zhang explain in *A Brief History of Qi*, there is a resemblance between prana and the Chinese notion of qi (vital force), the Zen notion of *ku* (emptiness, sky, space, air), and the Ancient Greek idea of pneuma (life spirit).

Ancient Greeks described a concept which in several important aspects parallel the Chinese notion of *qi*, with the word "*pneuma*." Like the Chinese *qi*, this Greek word is often translated into English as "breath"—with similarly misleading results. The Greek *pneuma*, like the Chinese concept of *qi*, was a complex idea that blended spiritual and material aspects of the vital essence of life into a comprehensive description of that without which life could not exist.

Early Buddhist zazen practice, Ancient Hindu yoga, and Ancient Chinese practices of qi all share an idea of the world held up by breath, of an initial unity that is like a breathless breath, or breath that breathes itself. Breath is untainted by heat, desire, thought, and these spiritual practices search for something beyond subject-object and mind-body distinctions, moving towards this original life force.

Modern yoga, on the other hand, is part of a martial tradition traced back to the training of ascetic mercenaries for battle in the sixteenth and seventeenth centuries, and its athletic version was born probably in the last 150 years. Yoga has now been revived by the Indian military, and at various points in its history has been used as training for orphaned children to keep them off drugs or out of other forms of trouble. This is hardly the image one has of enlightenment sold by the world of wellbeing. Like psychoanalysis, modern yoga was developed during wartime and has been imbricated in the study of the ways the individual is affected by violence and modern society. It has not been, at least for some time, simply an access route to the sacred.

I once had a yoga teacher who used her ten-minute opener to speak about her bad Tinder dates. She was okay as a yoga teacher; I was generally amused by the situation. I listened to her speak about the importance of yoga—how it helped her remain centered. She was waiting for the universe to open itself to these toxic men who need to find serenity. "I focus on my breath after I get home and think about coming to practice with y'all on my Sunday morning. Let's start on our backs!" Later, I switched to ashtanga at the urging of a partner who said the yoga I did was lame. I remember telling a fellow psychoanalyst about this, and she asked me if he was traumatized. "Ashtangi are the most traumatized people I've ever met," she said. "It's the only thing holding them together. Like a carapace. Really, they are all reformed drug addicts there." I laughed.

Yoga is an easy target. A bit like Freud. But I really do admire the discipline that yoga requires and the attunement to the body it engenders. I went on a retreat in Goa with Western instructors who studied ashtanga yoga in Mysore with Pattabhi Jois. A rather depressed assistant told me that I would never achieve *Mula Bhanda*, or root lock strength, because I had had a child. "Your pelvic floor is a mess," he said, removing my printout of the postures with a glare: "You must memorize them."

Ashtanga is the more extreme form of yoga. I stayed with it because its instructors are not only rigorous but mean. There isn't a lot of talk about being kind or gentle to oneself before class. It doesn't come with the life lessons one receives without consent. There isn't much exchange of personal anecdotes (though there is a lot of nostalgic talk of the early days with Jois). It is, in other words, as close as one can get to a yogic equivalent of Freudian psychoanalysis: daily, strict, orthodox, mute. The practice should break down your defenses and rebuild you.

My instructor wasn't wrong. I have a weak *Mula Bhanda*, though it has markedly improved since those damning comments. He also told me he worshipped the Hindu goddess Kali who represented the darker side of enlightenment—the Jungian shadow self. Kali was death and destruction. He showed me pictures of her with knives stepping on mutilated bodies. I wasn't entirely sure what he was communicating to me, though he indicated that there was precious knowledge he had that I perhaps might want but that I also wasn't ready for; maybe I would never be ready, no *Mula Bhanda* and all that. He eventually talked about a pilgrimage he wanted to take to a Kali temple but couldn't afford. I transferred him some money.

I wondered what his decree about the possibilities sheared off by pregnancy meant for women in general in the land of ashtanga. That a good woman like Kali was on the side of death not birth? Would women never achieve what men can in physical and spiritual strength? What does this mean for the yoga practitioners in the West who predominantly identify as women? A whole new army of the inadequate?

Ashtanga, in fact, has long been criticized for a misogyny implicit in its practices—like not being able to do yoga while menstruating. This is also linked to the reason you do not practice ashtanga on moon days—during the full or new moon—which is thought to affect your personal energy. You are also not to practice during early pregnancy, nor do many postures in late pregnancy. In fact ashtanga was developed for men only. But what distinguishes ashtanga from other yoga practices is the way they force you into postures—often quite violently. The adjustments are severe. This was always the appeal for its practitioners.

Ashtanga is practiced six days a week and the sequences take from an hour and a half to two hours. The teacher tells you once you have advanced from series one to series two and three; very few make it to the third. This is real devotion. The athletic dexterity of its most serious devotees is mind-bending, exactly the kind of gymnastic feats one imagines when one thinks of yoga. Not the loose flailing about that takes place in the average class. Over time the relationship between breath, strength, flexibility, and endurance creates a terrain of knowledge about the specificities of your body. Achieving strength and dexterity felt miraculous. I can do things with my body I never thought would be possible, and I still can't do more than half of what the best practitioners are able. And I can breathe. I know how to breathe.

Still, yoga does begin to resemble an addiction. Your body craves it. The ambition is a powerful motivator, as is the calm that follows that seems to break your ambition. I wonder if the birth of my daughter was so much easier than the first because I had gained control of my pelvic floor. Then, second births, they say, are always easier. Who knows? But what yoga has decisively done is keep my back from falling apart due to sitting all day listening to patients. Full-time private practice will take its toll. I don't think anyone is meant to sit for that long with that kind of person-to-person intensity; I was too young to be in that much pain. There was very little that Western medicine could bring to the problem aside from pills. As so many do—especially women—I turned to Eastern healing.

I had watched the mostly geriatric board of the psychoanalytic institute I belonged to bleed its members to back surgery. These are difficult surgeries that are a kind of last resort. Keeping muscles strong and joints flexible is important as we age. My chiropractor—a real 1960s man who spoke with a thick New York accent, had a white-haired ponytail, and moved like he was on cocaine—told me that no one is in more trouble than the shrinks. "You sit and strain under emotional tension. All their heads are popping off! Come to see me once a season. Once in the summer, once in the fall, once in the winter, once in the spring. Take care of your spine like your teeth."

The solution to chronic back pain is arduous to achieve. I needed to build up my core muscles so they would hold my back; develop a different relationship between my tailbone, pelvis, and rib cage, which were either splayed out or curling under; and work with my hip flexors, which were overstretched from the posture of sitting, scrunching them in the front and pulling the muscle in the back. In fact, I needed a carapace because I was hyperflexible and weak: a serious matter but rather typical of what sitting achieves and what yoga works to undo.

Ashtanga taught me how to breathe. The effects are undeniable. *Ujjayi pranayama* is necessary to get into a posture, no less endure a long sequence of them. "Breath of victory" is the common translation. You constrict the back of the throat, allowing you to slow, control, and extend your breath. It sounds like the ocean. A room full of people doing *Ujjayi* is hypnotic. All the benefits of nose breathing are brought, little by little, back into the life of the pathetic, unfit mouth breather. It is a reminder to see how shallow you breathe, moving air into "dead space" in the lungs, the space that isn't being used.

It took time not to feel like an absolute imposter. Having been told for years to breathe into my hips, or breathe into my feet, instructions that felt quasi-mystical when not absurd, things finally started to make sense. As did many of the other instructions by gifted teachers who can speak you into your body. Only now could I really feel the effects. Over time yoga cured the tension in my jaw and stopped me grinding my teeth. But to get there takes time, experience, attention, and, most of all, repetition. Four to five times a week is critical. As the psychoanalyst in me says: you have to spend more days doing it than not for it to really work.

Importantly, the idea of ashtanga is that it becomes *your* practice. Eventually, you don't rely on a teacher. Classes are not led by instructors who are the only ones who know the sequences. It is supposed to be a self-limiting commodity. The transference to teachers will be overturned by developing your knowledge of your body. Ashtanga is the one yoga practice that truly made this possible by taking away from one's vision a teacher to follow.

Transference makes you work. Transference is a moment of falling in love and wanting to see where this love will take you. Transference is also a form of insanity. I had one instructor who was the second-hand man to his girlfriend, who was the more famous teacher. He was finally allowed to lead the class (these happen once a week). When he started announcing the postures and counting for the holds, he did so with a fake Indian accent. It took all my strength not to break out in laughter.

Wasn't this the moment of truth? Was I not in the middle of an unanalyzed transference to an unassimilable foreign country? Was there any point at which they would dissolve this madness? Probably those that did analyze it left the fold with *their* practice in hand, just as a good psychoanalysis terminates. As Winnicott said about patients, the capacity to be alone is a real milestone.

Breath work and psychoanalytic work are two modes of learning how to breathe. From a certain angle they are diametrically opposed. One is about dropping thoughts and being closer to the body—staying there for longer and longer periods of time. The other is about dropping down in your thoughts to the bodily edge of consciousness but staying within a world of representation. But similarly to how Bion imagined the psycho-somatic and the soma-psychotic, perhaps with these two practices we are simply looking at the question of breath from both sides.

When I think about yoga, I see these beautiful forms composed of body and breath. The forms are held and then dissolve as one moves into and out of them, sometimes slowly, sometimes very quickly. It begins to look like writing with your body. "Whoever writes poetry engraves forms in our memory," the Austrian poet Ingeborg Bachmann once wrote. "And I believe that whoever inscribes these forms also disappears into them with his own breath, which he offers as the unrequited proof of these forms' truth." I have always felt that psychoanalysis is closer to poetry than to narrative—especially if narrative is used in service of explanations. In psychoanalysis at its best, the basic forms and sounds of language feel fluid and ephemeral, never quite solidifying into anything like *my* story, single and permanent.

ASPHYXIATION

SINCE THE PANDEMIC, I've felt as though life has become unhinged in a way that is hard to put a finger on. Patients are caught in spirals of anxiety and morbid thoughts about the future, with no exit. I understand what it means to be in rut. This feels different and differently collective, as if we were all spinning in the same vacuum. We charge ahead with our personal and professional pursuits with abandon, while wearing ourselves down and fighting a feeling of the meaninglessness of it all. Political debates feel violent, locked into messy tribalism with little that resembles dialogue or the step-wise unfolding of perspectives. How can we even begin to care?

I remember speaking about this with my colleague Eyal Rozmarin. We were discussing how clinical work had come to feel so odd, as if everyone was subtly off balance. I remember his words so clearly that afternoon, as we walked in Brooklyn: "We lost all the rationalizations we use to keep submitting to the machinery of life. One rationalization building on the next and then the next, and then kaboom! We don't quite know the reason we do anything anymore." I almost jumped. I immediately knew he was right, as if we were all living in the aftershocks of an earthquake. None of it is really holding together anymore—as if we were all pantomiming normality. But the seeds of doubt had been sown.

We tend to forget that we're breathing. During the pandemic, this forgetfulness became untenable, as breath, the vector of that plague, was driven from background to foreground. I can't help but notice the turmoil that seemed to arrive at the very moment that breathing was thrown into question—not just the effect of coronavirus on the lungs, but the political situation in the United States, our asphyxiating history of racism and police violence, climate catastrophe, all amid renewed international wars. It's as if the body and environmental threats were directly translated into the political arena and made explicit.

Many like to speak about the pandemic as an era of anxiety, much like that of the twentieth century and the world wars. But one also hears talk about this moment being different, moving so far beyond any individual angst or national agitation. It's not only that we don't know why we do what we do, like answering fifty emails a day, or spending countless hours on social media, or trying to save money, or watching the news, or even protesting. We also don't know what to care about or what makes a difference. There is a decidedly apocalyptic sensibility at play. It's as if we can't make sense of life anymore, no less of the future. As Judith Butler titled their book on the pandemic—*What World Is This?*

It wasn't only breathing but also social conditions that the pandemic brought crashing to the fore. Never had we been forced to confront the stark opposition between the pursuit of individual freedom as America has so gloriously and so vainly defined it and the reality of 9 billion inhabitants living and breathing in a shared atmosphere. As our breath became harmful to the other, restriction became the essence of collective care. Maybe it was always the essence?

If the pandemic has undone something, then perhaps it is our feeling of being bound together by a human project. The philosopher Jean-Luc Nancy called coronavirus a "communovirus," pointing to the communal aspects of life unearthed by the pandemic, along with an ironic nod to the renewed fears of China. Even as it showed us as equal participants, we understood that we are not all equal in our privileges when it comes to life and death— just as, even outside of times of pandemic, we aren't equal when it comes to the air we breathe.

So many public intellectuals hoped that the break produced by the virus would lead to new forms of consciousness, only to have their spirits dashed by the brutal New Normal that followed in its wake. Economic forces cashed in on our all-too-human fear that spread as quickly as new variants. The communovirus didn't make us more communal; it seemingly made us less so. Less likely to care for one another and more focused on protecting our own.

My most powerful memories of the pandemic are from the long hours I spent in the intensive care unit during the first wave. Often, I was facilitating painful, one-way conversations between loved ones and their intubated family member that they could not visit. The survival odds for an intubated patient during the first wave were slim. These patients were likely going to die.

One day at the ICU I helped a man see his father on a ventilator using WhatsApp on my phone since he couldn't figure out how to use the Zoom system on the hospital tablets. The son was in Haiti. He had been told his father was "crashing," meaning his lungs were giving out. It is a word I heard constantly. Computers crash, stock markets crash, cars crash. Now lungs were crashing too.

The sound wasn't working on my phone, but his father was in a medically induced coma, so the man said it didn't matter to him—he just wanted to see him. The internet connection wouldn't hold the video call for more than fifteen seconds, so we figured out quickly that he could just call back and I would flip the camera each time so he could see his father. We did this a dozen times before he called me and said, "Okay, that's enough." It was a moment of strange and unexpected intimacy: not merely being present at this moment but seeing him seeing his father on the double screen of these internet video calls.

These intensive care units were makeshift wards crafted out of corrugated plastic barriers, separated from the main areas of the hospital by interim stations where you could put on or take off your personal protective equipment. For a time, I didn't think about the risks. Until I saw too many people expire before my eyes and I started to imagine the infection on my hands, my phone, my shoes. All the moments of failing to wash properly. The reality of death finally got to me. And it cracked a façade where I had been willing to continue working as a volunteer in a hospital that certainly was not simply or only a humanitarian enterprise. It was a business that willingly jeopardized its workers, no less its volunteers.

Who can forget those quasi-experimental scenes of a person sneezing or coughing and the neon-green cloud of microparticles wafting out of their faces hundreds of feet? They seemed endlessly repeated on the news in those first months, especially as the question of mandatory masking laws were debated with ferocity. The hospital was covered in masking tape that was green, red, yellow to demarcate areas with different risk levels. Of course given how germs travel, none of this partitioning made a huge amount of sense (though it must be said, I never got sick once in all those months of working), but as with traffic lights, which don't necessarily reduce accidents, the real purpose of this system was to give us a semblance of order and safety, important in this scenario where staff also fell sick and died.

In the red zones you had to wear a gown, gloves, face shield, N95 mask, and hair and shoe coverings, much of which would get crammed into garbage bins in the yellow zones, before walking out into the green areas. I religiously washed my face shield before exiting the hospital and putting it in my bag; we had to reuse them as they were in short supply. When I came home, I undressed on my doorstep. I left my coat and sneakers outside before putting my clothes in the laundry and taking a shower. It's incredible how quickly you can develop a routine, a rhythm, to hold yourself together. It was as if all of us adapted to a new order within a few weeks' time.

Most of the hospital was a yellow zone, except for the entrance and the staff cafeteria on the first floor. Patients' rooms had large red signs on their doors indicating an active infection. But these rooms were entered and exited via the hallway where staff congregated. You would put on a fresh gown before walking into a patient's room and then discard it when leaving. This system would become somewhat delirious if you had to visit many rooms in succession, putting on and discarding gown after gown so as not to bring too much infection into the yellow zone.

Patients were also being transported from one place to another. Hence the diminishing PPE. The trashcans continually overflowed with yellow and blue gowns and masks, which hardly added to a sense of safety. All this encapsulated in miniature what was happening in the world at large: the impossibility of drawing absolute boundaries in a communal atmosphere.

Even in this context, the ICU was an odd place. On the one hand, it was the very heart of the red zone, containing those who were closest to death. On the other, it was actually much safer than other wards, since everyone there was comatose and breathing via machine, unlike the wards filled with patients who were talking and sneezing and coughing.

We worked in silence in the ICU. It was an unspoken rule that if you wanted to converse you went back into the hallway areas. Who was this rule for? The patients wouldn't have been disturbed by the sounds of talking. Was it like holding your breath when you pass by a cemetery? The intensive care nurses—a special breed—seemed to float as they tended to the bodies, moving from bed to bed. It almost felt like we were all already dead. Weirdly I found peace in the ICU. "To pray is to breathe," says Kierkegaard. For the first time in my life, I prayed, not for myself but for these people, reduced to bodies, living by machines.

It is difficult to understand where the person is when they are in a coma. You begin to play a game with yourself imagining they are there, just below the surface. You muster all the faith in the world that somewhere they hear you and know you are with them. Lacan says there are only a few rare experiences in a life where we touch what he called the Real—that aspect of reality that falls outside of the usual ways of organizing experience and our sense of self through images and language. It isn't often that we truly know ourselves as living flesh in a process of dying. This can certainly induce trauma, as became clear from the heated atmosphere on the ward, particularly when things were calmer, so that the pent-up fear, anxiety, and rage of the exhausted staff was finally given space to breathe.

Surrounded by all those dying bodies, I watched a whole world of belief blossom in my mind. Jacques Lacan felt we have very paltry ideas about our own death; far more is at stake than life itself, which we, in fact, know very little of. For Lacan, it's always a question about living with death, living with loss, living with limitations. Once you are dead, you don't have to live with any of it.

Freud is even more exacting on this point. In *The Future of an Illusion* he speaks to an imaginary interlocutor about religious beliefs against those of science:

We desire the same things, but you are more impatient, more exacting, and—why should I not say it?—more self-seeking than I and those on my side. You would have the state of bliss begin directly after death; you expect the impossible from it and you will not surrender the claims of the individual.

"I want to be in this solemnity of the dying," I remember thinking to myself. Looking back, I can see this was just another defensive maneuver. In contrast to the other wards, the ICU had become for me a place of refuge, an almost sacred space whose silence was only interrupted by the sound of breathing machines. I wanted to be closer to death precisely to get away from the sick. By choosing to be here, I was trying to get away from the patients anxiously struggling to breathe—patients who were terrified of being intubated and taken *there*. They would be dragged across the threshold and never know it; calmly and defiantly, I crossed it daily.

Just when one thinks one is confronting death at its most stark, there are still layers of illusion. The philosopher Havi Carel writes that breathlessness brings to the surface a feeling of bodily doubt in people who usually experience a tacit certainty with respect to their bodies: "Breathlessness is a core example of this breakdown. In cases of pathological breathlessness there is a break from a past ability to exert oneself and to perform actions freely." Many aspects of illness compromise free action, but breathlessness is unique since it must be constantly negotiated, breath by breath. This leaves the other who witnesses this negotiation likewise helpless. This "tacit certainty" may itself be an illusion, but it is one we all like to walk around with. It gives us an aura of safety.

Breathing difficulties shatter this supposed certainty. And global breathing difficulties seem to have shattered our greater "tacit" certainties, especially as communally shared certainties.

Looking back, I can't help but wonder if, on some level, I volunteered at the ICU as a way to preserve my sense of freedom. Elsewhere at this time, the Western obsession with agency was showing its true face: the right to liberty was defiantly expressed across the United States and the United Kingdom as a right to get sick and make others sick, the right to spread death at will, as if willfully spreading death was the very essence of our entirely negative conception of freedom as freedom from anything compulsory. In the United States, government was there to protect this freedom unto death: "My freedom doesn't end where your fear begins."

Before COVID, I never really understood why psychoanalysis seems to have such disdain for the idea of freedom. It felt like an affectation particular to the French intelligentsia: "*Je ne parle jamais de liberté,*" Lacan once declared. Since the pandemic, I think I have come to see what Lacan means, especially when he said that freedom's most faithful companion is madness. I can't help but think of the signs I'd see angry protestors waving on the television in the lobby of the hospital: NO MASK MANDATES! GIVE ME LIBERTY OR GIVE ME DEATH!

COVID has shown us the limits of freedom, conceived narrowly as *my* freedom. What does that mean when we all breathe the same air, share the same spaces? As Lacan explains, the will to show a kind of unity—either as a united self or as a united group—often involves isolating a victim so as to feel a "bitter, jubilatory satisfaction" over them, or what he calls the "suicidal aggression of narcissism." Libtards. Snowflakes. Maskers.

Lacan quotes from Molière's *The Misanthrope*: "I'll flee this bitter world . . . And seek some spot unpeopled and apart. Where I'll be free to have an honest heart." Freedom, understood as a private luxury, is always misanthropic: I want a space that isn't a space of shared breathing! Lacan is less interested in liberty than in how interconnected and interdependent we are. It would be better, he argues, if we were interested in truth over freedom, and thus attended to the problems of establishing truth, especially truth that can apply to more than one person, one life.

"We cannot really live without each other, without finding ourselves inside another's pores, or without letting another in. For that is where we live, outside of the bounded self and its conceits, as an opening toward the world," writes Judith Butler in *What World Is This?* In making vividly apparent the fact that we live in a shared atmosphere, the pandemic gave lie to the illusion of the individual who is free to act, entirely separate from others and its environment. Never before had it been made so manifest how radically porous we are to the air that others breathe, to the fact that they move through us. "A holey body is rarely accepted socially or politically," writes the geographer Marijn Nieuwenhuis. Porousness "triggers emotions and feelings of discomfort, pain and revulsion even though our body leaks all the time."

In the ICU I saw a man don a prayer shawl and recite Psalm 115 to his father:

Why do the nations say, "Where is their God?"
Our God is in heaven; he does whatever pleases him.
But their idols are silver and gold, made by the hands of men.
They have mouths, but cannot speak, eyes, but they cannot see;
they have ears, but cannot hear, noses, but they cannot smell;
they have hands, but cannot feel, feet, but they cannot walk; nor can they utter a sound with their throats.

This will have been the effect of coronavirus for me, and maybe for some others: to have wiped out the false idol of freedom. What will it be replaced with? I didn't know the Bible verse before hearing it in the ICU, only Shakespeare's rendition of it: "Last scene of all, / That ends this strange eventful history, / Is second childishness and mere oblivion; / Sans teeth, sans eyes, sans taste, sans everything." This is the only freedom that psychoanalysis raises up, namely the importance of speaking freely—say everything, say anything. Speak, even now, even in this last scene of all.

IN THE ICU there was a series of names on the IV equipment connected to the ventilators that indicated the medications and their dosages keeping the patients in their medically induced comas. One caught my eye: fentanyl.

Fentanyl is one hundred times more deadly than heroin, and highly addictive. In 2022 it was responsible for two hundred deaths each day, many of them because the drug suppresses the respiratory drive, leading to hypoxia. Another epidemic of death by breathlessness: more than a quarter of a million Americans have died from a fentanyl overdose since 2018. Most of these deaths took place outside a hospital due to the rapidity of the onset of the sedative effects of fentanyl and the difficulty of getting adequate medical attention in time.

In cases of fentanyl overdose, patients often need to be put on a ventilator, just as my patients in the ICU were. In fact, this is why COVID patients were given fentanyl: its respiratory depressing function was exploited to allow the ventilator to manage and control breathing. A

fent is a slit in fabric—a breathing hole. Close to *vent*. The word appears to already mean to stop breathing: *fent-nil*. A fent is also a piece of waste fabric, a remnant.

I've stared at the signs on the subway asking everyone to carry naloxone, the opioid antidote that can reverse an overdose in the form of a one-time nasal spray. The ads showing pills or bags of drugs asking if you can tell which one has a deadly dose of fentanyl: "We can't tell either." The ad showing a pushpin with a minuscule amount of white powder on the head of the needle: "This is enough Fentanyl to kill you." A colleague who works with the homeless told me she was taught to rub her knuckles on a person's chest to see if they are slipping into a coma. "Does it check their breathing?" I asked. "No," she said, "I think it just hurts a lot."

Knuckle was the word a woman used in the *New York Times* to describe her clitoris after taking SSRIs for a decade. She is among those who claim they have never recovered from the sexual numbness that came from taking depression medication. Meanwhile, psychiatrists insist they are just depressed from not taking their drugs. So, they would be numb with or without them? Depression is obviously a cause for suicide, but so is chronic sexual dysfunction. Depression causes sexual dysfunction, but so does its cure. An impossible loop— claustrophobic and closed—as if there were no exit from sedation.

Increased prevalence of fentanyl is related to big pharma's aggressive promotion of opioid use, creating, as just one example, the Sackler family's $11 billion OxyContin fortune. Fentanyl, unlike heroin, is cheap. Very cheap. And very, very strong. Prescription opioids are often the gateway to street variants, even more easily obtained than a prescription. In 2019, there were 1.6 million new users of prescription pain medication, the majority of whom said they took them for pain not for pleasure. These drug companies profit on human life. Only recently have they been shaken by protests and criminal prosecutions taken on by civilians before any governmental institutions would become involved in attempting to regulate the problem, or even care about it.

I don't think the concept of addiction should exclude suicidality—a will to self-destruction that is an attempt to master suffering and pain. Of course, many users don't know, or only half know, that the drugs they are taking contain deadly amounts of fentanyl. Yet knowledge of how fentanyl, with its unparalleled mortality, has contaminated the drug supply has done nothing to diminish use. In fact many people are now consciously using fentanyl, while those who survive overdose often speak about all their friends who have died. Listening, one begins to hear a will to join them, grief dragging them in the direction of death.

Beyond the language that usually surrounds addiction and personal responsibility, suicidal aggression congregates around the tears within a social fabric. As an epidemic, it reveals those places where people might be welcomed into an identification with what society fails to acknowledge and even rejects, throws out, seems willing to dispense with and reduce to waste. Fents, remnants. Fentanyl and opioid use can be seen as a socially undergirded attraction, even seduction, into a feeling of sedation, a withdrawal from life. Doesn't every suicide pose a question about the Other's desire for them to exist?

I usually hear from my patients the punitive voice of the law behind their depression. A condemnation that paradoxically permits them to abandon themselves to enjoyment while excoriating their very existence as something to enjoy. In fact, the two go hand in hand. We hear a lot about the fentanyl deaths among the LGBTQ homeless, the spate of deaths in impoverished towns whose former sources of employment have abandoned its population to a crisis of survival, as well as our distressed and fragile teenagers. The persecutorial voice is always a voice set against a person who is told they have no right to protect themselves either from unmitigated enjoyment or from punishment.

Sedation and the desire for death are close. Some might prefer the idea of passive suicide, but again this places the act on the side of the individual. We need to think about suicidality as a social phenomenon. What are these deaths telling us? What are we failing to hear that returns as an issue of breathing and pain and death?

In my count, then, there is a third epidemic at hand— suicide. Suicide rates in the United States increased 37 percent between 2000 and 2018 after having decreased significantly in the 1990s. There was a small decrease between 2018 and 2020, but rates returned to their peak in 2021, where there were an estimated 1.7 million attempts. More than 12 million adults say they have had serious thoughts of suicide. It is the second leading cause of death in the 10–14 and 20–34 age ranges. Suicides among the young, those aged 10–25, increased by more than 52 percent in the years between 2000 and 2021. These numbers are staggering.

One of the other startling data points on suicide is the increase in deaths by suffocation, which is more deadly than poison and cutting. In the United States firearms are, of course, the leading means of suicide, but the use of suffocation, especially among females, more than doubled between 2000 and 2020. The biggest increase was among those aged 10-24, where there was a more general decline in firearm-related suicide.

The increase in suicide by asphyxiation holds in other countries, from the UK to Australia (though their stricter gun laws in comparison to the US should be noted). From 2007 to 2018, 78 percent of suicides in children aged 5-11 were due to suffocation or hanging. Why is asphyxiation the increasing means of suicide for women and children? Why are we increasingly finding ourselves within a scene of death by asphyxiation?

This personal scene of death is taking place against a social backdrop of climate crisis—that is, a backdrop of collective death by toxic air, extreme temperatures, wildfires, and smoke inhalation, as well as death by drowning due to rising sea levels, or excess rain and floods. In the US, in particular, there is the continued violence against racial minorities, often in the form of insidious environmental racism, which means that people of color are more likely to breathe polluted air—the fallout of a long history of public lynching and slavery. During the pandemic, this violence extended into how COVID disproportionately affected racial minorities. This was palpably true in those first months that I was in the hospital, when most of my patients were Black and brown minorities and other immigrants.

In the summer of 2020, the politics of breathing took a sudden turn, from COVID to the issue of racism. The Black Lives Matter movement rallied around the slogan "I Can't Breathe," the last words spoken by Eric Garner, caught on video, while he was suffocated by a police officer, in 2014, and in 2020 uttered repeatedly by George Floyd over eight minutes while his neck was crushed under the knee of police officer Derek Chauvin. The murder of George Floyd set off a series of protests in cities across the United States.

Having stared at this word *fentanyl* on respiratory machines for weeks, I decided to leave the hospital for good, joining the protests in Brooklyn where I live. This was a moment when thousands of people suddenly broke lockdown protocols and reclaimed shared space, now as a space for political contest. This was well before we understood that breathing together outdoors was safe. It was an extraordinary moment.

Despite everything the pandemic brought to the surface, little has changed—police killings, fentanyl deaths, and suicides rates have increased since our return to "normal." Having ignored Achille Mbembe's 2020 call for a "universal right to breathe," we are facing the reality of an "I can't breathe" that is making an insidious appearance in too many aspects of contemporary life. Fentanyl, virus, breath, suffocation—these all feel like metonyms for the collapsing struggle to be counted as worthy of life. Amid all these injustices, how can we make a body count? A difficult question, but one thing I do know is that the way to regain control of breathing is to start counting.

DOMESTIC VIOLENCE INCREASED dramatically during COVID, as many women were confined to their houses with abusive partners (we are now at our fourth epidemic). Jacqueline Rose pointed to the fact that men confronting the specter of death took their rage out on women. Indeed, Freud himself saw how tyranny could be the secret companion of bodily panic, suggesting that where women became hysterical men became tyrannical. In his phantasmatic text *A Phylogenetic Fantasy: Overview of the Transference*, Freud imagines humans first confronting environmentally threatening conditions during the Ice Age, with men promising women magical protection in exchange for submission. When their magic runs out, there is no end to their violence.

Ingeborg Bachmann diagnosed a similar drift after the Second World War. Menaced by the encroachment of fascism into everyday life, especially domestic life, as if the war had never ended, she sought to write from this point of feminine "insanity." Her taste for sedatives, pain killers, and psychotropic medications blossomed in response to what she said was a subtle environment of infinitesimal crimes, especially against women.

Bachmann died after passing out from a massive dose of sedatives while smoking a cigarette in bed, which then caught fire. I don't know if the sedatives, smoke, or fire killed her. She would say it was the environment of crime. "There's no such thing as war and peace, there is only war," she writes. For Bachmann the idea of deliberate suicide was an anachronism—an idea arising out of a Western notion of a sovereign self who can take their life into their own hands. This is a notion only relevant to those who had the privilege of calling themselves a subject to begin with. No woman has the luxury of suicide.

Today, Bachmann says, one is always murdered. We live in an atmosphere of murderousness, so we can only slip into a death that was waiting in the wings with open arms. All suicides, she says, are now closer to murder. In an atmosphere this suffocating we are drawn into death. In an interview she once said:

> I've often wondered, and perhaps it has passed through your minds as well, just where the virus of crime escaped to—it cannot have simply disappeared from our world twenty years ago just because murder is no longer praised, desired, decorated with medals and promoted.

Likewise in our own era, everything is becoming too Real, in the Lacanian sense of shattering any feeling for sense or coherence. Even if coherence relies on an illusion, there are important structures that a society must provide its citizens. Of course, any instance will fail the test of coherence or sense, but the social fabric today has too many visible holes to fall through.

We can see this on the internet at hyperspeed: endless attacks and interpersonal misunderstanding in the form of comments and posts, Reddit subgroups trying to hold together the isolated and ill at ease, the proliferation of contradictions and lies on a national level that news can do little to ameliorate, and a whole world of mental health speak that fails to do anything but proliferate diagnoses. All this, yet nothing to do but watch. We are even incited to enjoy it—doom scroll! "The Real gives me asthma," wrote the anti-natalist Emil Cioran, and it is in this current environment that, according to birthrates, today's younger generation has collectively agreed that it doesn't want to pass on the trouble of being born.

A similar atmosphere to that outlined by Bachmann appeared in the short book *Terror from the Air* by Peter Sloterdijk, excerpted from his 2,500-page trilogy *Spheres*, in which Sloterdijk writes a history of Western philosophy from the perspective of space and air. Sloterdijk argues that one of the key concepts of the twentieth century is what he calls "atmoterrorism." Atmoterrorism forces something into the foreground—namely air—that had previously been a background given, thus ushering in new forms of violence and control.

Sloterdijk claims that the emergence of atmoterrorism can be marked to the day. On April 22, 1915, a newly formed German "gas regiment" launched its first large-scale operation by unleashing approximately 150 tons of chlorine gas on French–Canadian troops at a concentration high enough to cause damage to the lungs and extreme respiratory distress. Thanks to fortuitous winds, this yellow-greenish cloud made its way across enemy lines, leading to mass death.

Previously, wars were always battles of intentional hits, a matter of taking aim at one's opponent. What changed on April 22, 1915, is the targeting of the enemy's environment. Henceforth, the enemy's milieu will be made unlivable by attacking ecologically dependent vital functions from respiration, central nervous regulation, temperature, and radiation conditions. These tactics were prophetic of the degeneration of wars into our current situation of infinite combat between states and non-states or "terrorist" combatants: "The terror of our times consists in the emergence of a knowledge of modernized elimination that passes through a theory of environment, which enables the terrorist to understand their victim better than they understand themselves." As I write this I cannot help but think of the failures of the current war by Israel against Hamas to distinguish innocent Palestinian civilians. Walled into the Gaza Strip, they have been cut off from electricity, water, food, humanitarian and medical aid, and reasonable shelter from bombs, thus removing the conditions necessary for life.

States and terrorists now actively terrorize civilians and their environments, attacking what we are dependent on: "the breather," writes Sloterdijk, "becomes at once a victim and an unwilling accomplice in his own annihilation." With gas warfare and the terror tactics that followed in its vein, a new vigilance is required that is directed not towards anything easily locatable but towards one's entire environment. The so-called war on terror, according to

Sloterdijk, is the idiomatic confusion of the twenty-first century. Terror proliferates: in the war on terror, every attack, from whichever side, is seen as a "counterattack." All murder is suicide and vice versa.

Terror of and from the air. Science and knowledge develop at a rapid pace in response to this new status quo. From gas warfare, we developed gas chamber technology, gas masks, breathing apparatuses made for flying at high altitudes, ventilators, scuba equipment, and air conditioning. In these self-perpetuating technological developments, Sloterdijk sees a need to perceive and control the unperceivable—not only air but the whole surrounding environment, every-thing that one had previously taken for granted. Through a process he calls "explication," every aspect of the implicit life world is brought to our conscious attention.

COVID is perhaps the latest chapter of this long history of atmoterrorism. However intimate and domestic those first days of lockdown, COVID brought to the forefront the global nature of shared breathing space made possible through air travel and a will to inoculate billions of inhabi-tants on earth. Like gas warfare before it, COVID produced countless new experiments: governmental control of the masses, segregation of the sick, new border technology for the restriction and monitoring of travel, the production and administration and sale of vaccinations for an entire planet, and eventually a resurgence of international wars.

"It is the defenselessness of breathing, which I would like to talk about," declared Elias Canetti in a 1936 speech quoted by Sloterdijk. "One can hardly form too great a notion of it. To nothing is a man so open as to air." In this speech, Canetti discusses the way gas warfare wraps its enemy in a poison cloud, something which he likens to new states of political mass hypnosis and hallucination that, in the days of Hitler, seemed to be travelling through the air, and which, during the pandemic, took the form of conspiracy, misinformation, and hatred. Today air is perhaps the last thing that belongs to us all, yet in our current physical and ideological environments, we are in danger of being collectively poisoned.

More than anything, COVID has revealed the defenselessness of breathing. Indeed, if there is one thing that has been made explicit by the pandemic and the political atmosphere of the years since, it is a collective need for care and concern—for a return to intimacy. Intimacy is what Bachmann yearned for most. Her life seemed to prove to her the impossibility of transcending what she called a "personal abyss," at once banal and horrifying. As her character Undine says before drowning: "am under water . . . almost mute, almost still hearing the call. Come. Just once. Come."

The wish that we could come closer to one another—so necessary for care—is an intimacy we struggle to encounter and begin to reinvent. We want to move, communicate, and anchor ourselves in air, but we haven't yet found a way of doing so that would be tantamount to the shared protections offered to the soul that are no longer available to us. As Sloterdijk puts it, we need an "atmosphere ethics" where the good is the breathable: "The most fragile," he writes, "is the starting point of responsibility." What is the most defenseless must be defended. Any collective care must start here, by refusing to turn away from those in need of an intimate hand. Their call reaches our ears through the air we share and draws us in. Come!

A PSYCHOANALYST FRIEND once told me I had a certain power to make people dream, and that he wondered where this came from. "Someone," he said, "made *you* dream." Instinctively, I replied that my mother was gone in the Philippines, and my father was always gone, flying an airplane. "Ah," he said, "the little girl who dreams of these people flying through the air."

Hysterical symptoms always speak to the world from which a desire is born, where what is lacking makes imagination take flight. What is experienced passively and suffered is then actively deployed like a superpower. "Why am I never on the receiving end of myself?" a hysterical patient once asked me, a question that, taken at face value, sounds like the utmost narcissism but that really speaks to the universal problem of always becoming what one dreams of coming—the embodiment of a longed-for intimacy.

While desire creates a space that felt necessary for the individual (and they do this unconsciously, almost automatically), this shifting symptom always charts something powerful in the sociohistorical landscape. This is why Freud is not in the camp of individual diagnostics and psychopathology: symptoms give the lie to the mind-body or brain-and-body division. They place their sufferers within an intimate shared breathing space, showing where this space is in danger of fully collapsing.

This collapsing intimacy is what Sloterdijk is suggesting when he draws attention to the way we always exist in shared atmospheres, countering the modern notion of the separate and enclosed individual. The idea of the individual—this foreclosure of intimacy—seems to license violence against the air and the environment. This, more than anything else, is what it means to recall the defenselessness of air: the fact that, whether we like it or not, the air we breathe inescapably connects us with others.

My own life is not so far away from the militaristic takeover of the air. My father was twenty-one when he flew his first jet at Mach speeds for the marines. He never participated in a war, but this was the way he lifted himself out of poverty and saw the world. His father also fell from a 200-foot radio tower and died when my father was fifteen years old. Of course he wanted to fly. I have pictures, taken by the military, of his flight over Mount Fuji. I also have the local newspaper article showing a picture of the body of his fallen father on the ground indicating the point on the tower from which he fell.

In my child's imagination it was all very *Top Gun*. He then became a pilot in the heyday of the commercial airlines—when it was still a glamorous job—before Delta declared bankruptcy in 2005, gutting its pensions (my father never really recovered from this betrayal). I spent my life with pilots—these men of the air. What I understood intimately was the phantasmic omnipotence of pilots, these military men who believed in the value of control. In those early days especially, flying planes attracted certain Icarus-like characters who believed they could defy death and fly with the gods.

James Webster died of a rare form of leukemia whose degenerative effects on his nervous system gave him unbearable hypotension, causing vertigo every time he stood up. He couldn't walk more than a few feet without having to sit back down. For some years I wondered if his symptoms were psychosomatic. I was wrong. It took them a long time to figure out he had cancer, partially due to COVID and the widespread neglect of the old it produced. The air force pilot, having spent his life guiding airplanes at high speeds through the air, died in a state of dizziness.

When my father was on dialysis, he decided to quit and go into hospice. It was a precipitation of his death, a way for him to oversee it. Everyone celebrated what they saw as his steadfast and sober courage facing death. When he and I spoke, I tried to explain that this meant he was going to die in two weeks. "I'm just going to try it," he retorted. "You won't see me again," I said. He didn't answer. I wasn't able to see him until he was close to dying and couldn't recognize me any longer. I watched as his breathing became a death rattle—a terrifying and unforgettable sound caused by irregular breathing and mucous that cannot be cleared because its pilot is losing consciousness. I had to translate this death rattle into what it was: the sound of the last breaths of my father.

My father died with his estate in shambles: with all his bills and taxes unpaid, without his will being executed in the way he said it was supposed to be (including failing to name his grandchildren), without handing over the passwords to protect his wife, who needed access to the accounts. He had always run a tight ship, every file organized and labeled, every detail sewn up. He also complained that no one ever helped, even as he made this impossible. How had he left everything undone? The denial of death makes you dizzy. It also makes you fail to take care of others. I had to be with him in his death as I had lived with him—the only one who knew the extent of his delusion.

The philosopher and psychoanalyst Luce Irigaray makes a connection between our denial over our own deaths and the figure of birth, in particular the breathing that a mother's body does for the fetus—a prelude to life in shared air. In her book *The Forgetting of Air*, she connects the failure to think about air with male philosophers' fixation on death.

Like Hannah Arendt, who challenged her lover Martin Heidegger's notion of being-towards-death with a question about birth, Irigaray places the peripheral phenomena of birth, breathing, and the feminine at the very center of philosophy: "To air he owes his life's beginning, his birth and his death; on air, he nourishes himself; in air, he is housed; thanks to air, he can move about, can exercise a faculty for action, can manifest himself, can see and speak. But this aerial matter remains unthought by the philosopher." Irigaray describes how air always escapes appearance even as it surrounds and envelops us, revealing a fluidity in being.

Air is an excess, at once free and at the same time bound in a breath, in our blood, in a space, in our earth's atmosphere. Irigaray calls air an "irritatingly excessive resource," one that gives us the sense of the possible. Here we are close to the figure of the Buddha who laughs and breathes, as Irigaray says, with all his skin.

This characterization of what an open expanse might mean for us begins to touch on another paradox. Yes, our entry into the world is a first breath and a cry, but it is also an exit before we can understand what has been exited and lost—our home within our mothers who used to be able to provide us with everything. "In air," writes Irigaray, "life is, in the beginning, the boundless immensity of a mourning. In it the whole is lost." Thinking about your own death, which you will never know, elides another loss that we all, in fact, know intimately—even if we have forgotten it. Thus, our relation to air moves between melancholy and ecstasy.

Irigaray finds the forgetting of air just oh so masculine, and Western male philosophy a rather closed affair that concerns itself with itself—its reason, its speech, its questions of the One or ideal. Whereas women's philosophy, she suggests, tends towards questions of relation, what takes place between two. Likewise, the misogynistic hatred of women and of Mother Earth is founded on an impossible mourning that is present in the forgetting of air. The terror of birth, losing our home within our mothers, leads to sexism and the forgetting of what we share. The elevation of the individual first triumphant breath is propaganda.

This constellation of forgetting leads us to that defiant freedom that refuses to recognize that freedom is separation *as well as* connection to others. People are not merely an obstacle but the means to our independence. How can we mourn this image of ourselves as omnipotent, lone free agents? I kept returning to this image of agency during those months of working at the ICU, when the TV screen in the waiting room constantly showed the protests of antimaskers, chanting about not being afraid, certain they were free from risk and furious that they were being asked to look after one another. Agency is such a lonely affair.

Like my father, you can't ever really allow yourself to ask for help. So being asked *to* help will always bring to the surface so much disavowed rage and resentment of the other's helplessness. I realized too late how much I wanted to help him, how little space I was given to do so. Little wonder I became a psychoanalyst—a helping profession.

How afraid must these lone agents be, if not of dying then of all those other people with whom they share a nation, air, tax money: this is the togetherness in air, the forgetting of which allows for so much willful destruction of the environment. Freud's message concerning our basic misanthropy, our inability to "love thy neighbor," calls on us to question the status of Eros at a given historical moment. Eros is the binding force in the psyche strengthened by the experience of intimacy—however excessive or ecstatic, however much mourning might be provoked. Eros, he said, always caused mischief, but this mischief should be welcomed in comparison to the death drive.

I wonder how closely we might ally Eros and air. Air is a material medium that we share and move through, along with, take into our bodies. Irigaray points out that this senseless medium is also the medium of speech, which emerges from and returns into the senselessness of breath. Sense is always bound by the senseless—by mere oblivion—the same way thinking rises out of the unthought or forgotten. Air, for Irigaray, is both the means and the limit of sensemaking—a bound excess that might be one of the best definitions of life.

How many films or TV shows have a scene where CPR is given to someone ailing? CPR is done at a rate of one hundred chest compressions per minute to the beat of the Bee Gees' "Stayin' Alive": *ah, ah, ah, ah,* breathe into mouth, breathe into mouth. Repeated for as long as it takes for the patient to gasp, for their heart to resume, and for them to breathe on their own.

There is something magical in these images of those lifesaving breaths (only sometimes is the nose depicted as being closed)—hence "the kiss of life." As a little girl, I always wondered if someone would perform CPR on me one day. Maybe mouth to mouth would cure my asthma. Learning the technique in high school was a scene of intense sexual excitement, even though we were to perform it on others of the same sex (as if that mitigated any of the excitement). We watched one another breathe into each other's mouths, giggling at high volume. I wanted to be a lifeguard and save someone's life.

And yet despite all this, and despite the fact that in hospitals it remains a routine practice, CPR is a brutal procedure that fails to save 85 percent of those on whom it is performed. Writing for *The New Yorker*, Sunita Puri calls CPR "akin to assault," describing how the compressions can break bones, puncture lungs, damage the heart, rupture vessels, and even burn flesh. This is to say nothing of the fact that, if the procedure does succeed in restoring breathing and a heartbeat, many patients will incur brain damage.

How has CPR become so standard that in thirty-eight states students must learn the technique before graduating high school? Puri writes that she feels as if she isn't doing what she's supposed to do if she doesn't perform CPR after a cardiac event, even when she knows it's futile. You can't quite tell where the pressure is coming from—her own image of what it means to be useful and heroic, or the expectations that she feels around her that she performs this display of doctorly duty. Both seem influenced by movies, which tend to portray a sanitized, *Sleeping Beauty-*esque version of the kiss of life. COVID lifted some of the magical aura around CPR because doctors were forced to speak truthfully about it given the risk of contagion. Still, its use persists. It seems remarkable how phantasmatically prone Western medicine is—that it took a pandemic to even begin to undo decades of magical belief.

The magic of CPR returns us to the idea of God's animating breath, or the soul that escapes through the mouth. If religiosity hides behind a medical ritual here, it wouldn't be the first time. Breathing rituals figure strongly in religion and, as we have seen, in neurosis. In his late work *Moses and Monotheism*, Freud argues that we have retained animistic and religious ideas tied to breath:

> If we may rely upon the evidence of language, it was movement of the air that provided the prototype of intellectuality [*Geistigkeit*], for intellect [*Geist*] derives its name from a breath of wind—"*animus*," "spiritus," and the Hebrew "*ruach* (breath)"... Observation found the movement of air once again in men's breathing, which ceases when they die. To this day a dying man "breathes out his spirit [*Seele*]."

While breath is associated with the soul or mind, at the other end of the spectrum, it can also be linked to anality through flatus. Lacan even remarks that God's booming breath and voice should be recognized for what they are when taken together—anal wind. The addition of anality to magical breathing takes the sacred edifice of breath as spirit down a notch.

Likewise, "blowing smoke up someone's ass" is an idiom about insincerely inflating another person's ego with flattery. Close to ass-kissing. The idiom is tied to a literal act, a procedure that for many years was erroneously used to revive the drowned. Appropriated from Native Americans, who used tobacco smoke to treat many medical ailments, it was absorbed into Western medicine in the eighteenth century. In 1774, the Royal Humane Society published this little mnemonic rhyme:

> Tobacco glister [enema], breath and bleed.
> Keep warm and rub till you succeed.

And spare no pains for what you do;
May one day be repaid to you.

While the method fell out of favor as medical science progressed, coffee enemas have been making a reappearance among those looking for a rectal reawakening.

In a 1946 article called "Cosmic Consciousness in Catatonic States," psychoanalyst Robert Clark discusses a schizophrenic patient he calls Professor X, who wrote about anal breathing and cosmic consciousness, describing the former as a lost art:

> The human method of using nose and mouth alone is cause of loss of vigor and spiritual power. Anal in-breathing is the true mode of deep breathing; its very practice expands the soul, gives it power.

Professor X says diaphragmatic anal breathing is akin to a total breath. This underutilized capacity is what sets humans apart from animals. If we wanted to count this as the idiosyncratic musings of a crazy person, we shouldn't forget how many of us are on the search for the full-body total breath as a (at least to my mind) phantasmatic mistranslation of Eastern breath practices. Qigong ejaculation control is, in fact, thought of as anal or perineal breathing. And scientists have been doing tests to see if oxygen infused into a kind of paste and inserted into the rectum like a suppository could raise blood oxygen levels in the event of respiratory failure. A new form of anal respiration might be on the horizon.

Another noteworthy example of ritualistic magical breathing goes back to Freud and his famous case study of the wolfman, who, under the threat of a forced termination of his treatment, gave Freud all the keys to his neurosis going back to the age of two. Obsessive rituals around breathing were part of religious ceremonials he engaged in as a means of omnipotent control. When making the sign of the cross he had to breath in and exhale forcefully. Freud understands that in the wolfman's mother tongue, Russian, *breath* was the same word as *spirit* or *soul*. The wolfman was attempting to breathe in the holy spirit and exhale his blasphemous thoughts. "He was, however, also obliged to exhale when he saw beggars, or cripples, or ugly, old, or wretched-looking people; but he could think of no way of connecting this obsession with the spirits." Why?

Freud finally solves the mystery of having to exhale when he saw people in a wretched state after a dream led them back to the wolfman's mother taking him at age six to a sanitorium to see his father, who was in a wretched state. The wolfman took enormous pity on him but also must have been quite frightened. This provided the reason for the ritual: the image of the father that he *wants* to take in happens through ritualistic inhalation, and the image of his father he wants to rid himself of must be exhaled.

Freud points out that all this breathing is also an expression of the wolfman's erotic identification with the father, whose heavy breathing he remembers hearing in the infamous primal scene witnessed at age two. The father fucking his mother from behind, whose penis disappears into her, is the source-image of both phallic power and fears of castration. The wolfman must valorize the upright, powerful, heavy-breathing father, like the menacing and magnificent wolves in the tree outside the window in his dream that gave him his nickname.

The glaring wolves gave Freud the keys to his analysis. This image of his father protected him from a feeling of utter undoing—the father's penis disappearing, indeed the father disappearing into sickness and madness, as well as all his money, which disappeared after the war. In Freud's famous case, symptomatic breathing is intrinsically related to a denial of loss.

I think we are closer to both the ritualistic breathing exercises of our nouveaux therapies and the way magical breathing still enters into modern medicine and contemporary ideas about psychological health, well-being, and the soothing of our suffering. We are still trying to perform magic—evade death, evacuate evil, enhance compassion, and reassure ourselves of the presence of some godlike omnipotent figure. Breathing is sometimes touted as a cure-all in the manner of religion.

Far from being a return to religion, breathing shows us that religion has never left us—especially since we cannot escape the problem of loss. Even as we attend to all this magical breathing, as with CPR, we shield ourselves from brutal realities like the real question of our shared fate under pollution and climate catastrophe. How can breathing bring us towards this loss rather than away from it? It feels like it's right there, in the breath we lose with every exhalation.

The theme of loss is central in the cases psychoanalyst Bruce Ruddick analyzed in detail in a paper from 1963 titled "Colds and Respiratory Introjection." He wanted to look at the recurrence of colds that didn't seem to have an infectious source. With one patient, they seemed to appear at moments of longing and separation. The more unbearable these feelings, the worse her colds. Ruddick even went so far as to detail one session for all the repetitive and symptomatic sniffs and coughs so that he could track the material they were connected to:

1. Discussing girlfriends leaving home—sniff and cough.
2. "At periods I could withdraw and for days would not see my friends"—sniff and cough.
3. "I can't stand women talking about nothing but their children"—sniff and cough.
4. The patient coughed entering a silence during which she had been thinking of a boyfriend who left without saying good-bye to her and she had felt hurt.
5. "Your window looks like a prison window"—sniff.

It goes on and on like this. It reads like a Beckett play! What a peculiar thing psychoanalysis was in its heyday.

I can't deny that I've thought a lot about the excessive sniffing of certain patients. Battles with colds and coughs, endless runny noses—these things are a large part of the frustration and helplessness of childhood. Freud's trope of displacement upwards speaks to moving sexual frustration towards something more familiar and manageable by moving from genitals to the mouth and nose. I once wrote about a patient's relation to snot, which was her childhood nickname and the first thing she thought when she saw a man ejaculate for the first time—like he sneezed on himself. This patient's hyperfocus on snot stood out against her own feelings of inadequacy, especially as a woman. This is where loss comes in, since all arousal and longing is predicated on the object not being there—hence its status as desired—along with the recognition that it may one day leave us for good (sniff, sniff).

In cases like this, psychoanalysis aims to restore the feeling of being able to find space for oneself. This follows the scientist and theorist Gaston Bachelard's beautiful and basic intuition in *Air and Dreams* that images of air (blue skies, clouds, wind, but also falls into chasms, the sea as a mirror of the sky, abyssal spaces) speak to imagination as extension in space, equivalent to an expanding sense of self. In art and literature the imagination is often figured as a rising in the air and a descent into the depths, close to breathing in and breathing out.

What Bachelard overlooks is that imagination is a move towards, not away, from others. Lacan once remarked that all analysts know the moment in analysis when, miraculously, the patient stops getting common colds. Of course Lacan is not suggesting that the cold virus is produced in the mind. But speaking towards loss, allowing a space for arousal and desire, is a necessary step towards the immunizing effects of being able to breathe more freely. Through analysis the neurotic can be slowly brought out into the open air they so fear, ultimately learning to embrace it. Here we begin to understand the threat and the thrill of being invited to "say anything" and to really imagine for oneself another life; this extension of the self towards others isn't for the faint of heart.

In *Bubbles*, Sloterdijk describes the scene of the fascinated infant watching a bubble drift away through the air. "While exhaled air usually vanishes without a trace, the breath encased in these orbs is granted a momentary afterlife. While the bubbles move through space, their creator is truly outside himself—with them and in them." Bubbles are fragile—even as the image of the sphere bears something of the promise of totality: little wonder they offer a vessel for our fantasies about the self. In the bubble we have the image of a first externalization of the self in breath. Then, pop!

It's as though we're always at risk of going too far with this extension of the self into the air. What is the difference between breathing that acknowledges loss and breathing that phantasmatically denies it?

Wilhelm Reich will be our cautionary tale. Reich was a disciple of Freud who emigrated to the United States during the Second World War, having been kicked out of the International Psychoanalytic Association (the only other analyst to Jacques Lacan) for his ties to communism. While Freud was certainly responsible for the excommunication of his early disciples (notably Carl Jung, Alfred Adler, and Otto Rank), Reich and Lacan were more officially banished by the professional association established in Freud's name. The "International," as it is often referred to, carried on the work of insuring psychoanalytic orthodoxy. As a profession, I think we are still trying to recover from these constant professional wars.

In Reich's 1945 book *Listen, Little Man!* Reich spoke about people's wish to remain sick: the little man cannot take responsibility for his life or pleasure, wishing to crucify and stone the great men who he refuses to see have brought him the truest benefits in his own life. Written twelve years before his death, Reich was seemingly already talking about himself as a martyr. In subsequent years, his ideas of sexual revolution would make him an influential figure among the Beat generation.

If there is a psychoanalyst of the air, it is Reich. His experiments in psychoanalysis led him to postulate a form of life energy—orgone—that he thought could be accumulated in a machine he invented to cure neurosis, sexual impotence, fascism, and eventually cancer. Later, this energy was shot out of large gun-like machines, a practice he called "cloud busting," in the hopes of changing weather patterns and chasing away UFOs.

"I still dream of Orgonon," sings Kate Bush in her song "Cloudbusting." "I wake up crying." The song is an ode to the psychoanalyst told from the perspective of his son, Peter, watching his father's tragic life unfold. Peter and his father used to go cloud busting at Orgonon, their home in

Maine. Reich was loved by the local farmers to whom he would promise rain when they needed it.

"Organon" was a family home, a cult, and a laboratory. Peter saw his father taken away by the FBI on charges of medical fraudulence. As Bush sings of this moment: "You looked too small in their big black car. To be a threat to the men in power . . . I can't hide you, from the government." Reich lost in a torturous trial, in which he attempted to represent himself, eventually dying in prison. His books were burned along with all his research papers and the stockpile of orgone accumulators.

From the eyes of a child, there must have been something marvelous in Reich's experiment. Even in my estimation, there is something powerful about Reich's desire to tap into the source of world discontent. Throughout my twenties, I kept a copy of *Listen, Little Man!* in my bathroom. What I find so moving in Bush's lyrics is imagining Reich, his son, and so many of his followers, waiting for the fruition of his work: a universal liberation—a great intake of fresh air—that would spell the end of fascism and war and tyranny and bad sex.

Orgone was the greatest promise made by a psychoanalyst, a promise that flooded an entire generation in America. Most of the West Village was under the spell of Reich. William Burroughs claimed to use his orgone accumulator throughout his life. Reich even gained the attention of Albert Einstein, who tested his accumulator box for changes in the energic field to no avail.

Others were skeptical. In his essay "The New Lost Generation," James Baldwin wrote that the Reichians had turned from an idea of the world being made better by political commitment to an idea of psychological and sexual healing. They reminded him of devout sinners at a revival running down the aisle. Indeed, Reich's tumultuous and violent temper in his relationships with women, especially when drunk, as well as maniacally controlling relationships with his followers, seems to have taken the path of so many cults during this period. One wants to admire—I want to. But, as Baldwin notes:

> They had not become more generous but less, not more open but more closed . . . There are no formulas for the improvement of the private, or any other, life— certainly not the formula of more and better orgasms. (Who decides?) The people I had been raised among had orgasms all the time, and still chopped each other up with razors on Saturday nights.

In 1933's *Character Analysis*, still a staple of many psychoanalytic trainings, Reich posited an emotional plague at work in civilization, "an organism whose natural mobility has been continually thwarted from birth develops artificial forms of movement. It limps or walks on crutches, in the same way a man goes through life on the crutches of the emotional plague when the natural self-regulating life expressions are suppressed from birth." This plague can reach epidemic proportions, as it did during the Inquisition or, as Reich would argue in 1946, during the years of Nazi fascism backed by Protestant sexual repression.

Confronted with such psycho-political deformations, the talking cure couldn't go far enough. So began Reich's quest to get at the source of the problem. In *Adventures*

in the Orgasmatron, the writer Christopher Turner writes
of his experience of Reichian somatic therapy with one
of his followers, Alexander Lowen, who instructed him
to perform various difficult physical feats—and most
importantly, to breathe. "Can you hear your breathing?
Well, that's what life is about and that's what therapy is
about . . . you see, Reich thought that breathing gives
your body life and if you do enough breathing your
emotions come alive and suddenly you're crying and
talking. BREATHE . . . otherwise you are half mechan-
ical." Heavy breathing, repetitive physical movements,
and the manual release of muscle tension would create
convulsions and vibrations in the body that would in
turn allow for stronger orgasms, help along the curative
spasming, and eventually cure disease. Lowen claimed
that his orgasms became like power balls inside of him
that shot into the stars, connecting with their energy
source.

The Reichian idea was to tap directly into soma and
energy, to return us to our most primal sexual selves, thus
bypassing any need for speaking—a far cry from say anything, say everything. For Lacan this fantasy is naïve at
best, and violent at worst. Norman Mailer, a Reich devotee,
put things a different way; this new world, he said, was
"the land of fuck." Reich's orgone energy was suspected of
being hidden in the body and, paradoxically, also in the
air. So began the dream of the perfect orgasm, the relaxed
and pliable body, breath freed from any tension, the just-
wet-enough atmospheric conditions, and the obliteration
of all enemies, domestic and alien. Mailer later admitted
that the apocalyptic orgasm had always eluded him.

For Reich this was not intended to be confined to the individual. At a global level it would usher in a new politics: "The removal of blockages from the sick body became a metaphor for the cleansing from the body politic of all barriers to freedom," writes Christopher Turner, who describes how Reich "would invent ever more intricate devices to combat the spinning forces conjured by his own mind."

Many were skeptical of the politics that would be practiced in this land of fuck. "This administration of happiness is nauseating to me," the philosopher Herbert Marcuse said in a conversation about the new lost generation that Baldwin was writing about that sought psychological and sexual well-being. Indeed, the pursuit of freedom and satisfaction was a new kind of hell. It made little sense, Marcuse said, to speak of excessive repression any longer—unless, that is, one suspected it of hiding within the new liberatory motifs. In this sense Reich is prophetic of our current worldview that incites us to enjoy and enjoy. In a perverse twist, everyone now feels they are falling short of the enjoyment they could and should be having. When you fail to enjoy, you must blame and punish yourself. Where is the room for loss in this will to enjoy?

This goal of total enjoyment, it must be said, is rather lacking in a sense of irony or humor, born from a sense for human frailty and contradiction, sexual mishaps and bungled intimacies, and most importantly the inescapable pains of life. There isn't even the veneration in Reich for the majestic construction of dreams or jokes—it's all unblocking and unlocking and accessing and enjoying.

Freud was skeptical of any wish to reach back before civilization, however rotten it is. We need to keep working on it, as on ourselves—developing our capacity for intimacies. How else can we do that than through language and the flawed institutions that we create through our representational capacities?

Kate Bush ends her song "Cloudbusting" imagining a young Peter Reich putting his father to rest: "You're like my yo-yo, that glowed in the dark. What makes it special, makes it dangerous. So I bury it. And forget." Where did all this revolutionary fervor go? When Bush made a music video for the song, she played the small-framed Peter to a much larger Reich with his even larger cloud busting gun. She interrupts the song suddenly with a repetitive chorus, panting "It's you and me, Daddy," like the secret call for an Oedipal victory, uniting parent and child in a cosmic-orgasmic battle. The clouds rally behind the two of them as they make rain as one. Then Reich is gone. Peter is left alone with the memory of his father and his tortured legacy. I think we are all reeling from the failures of the sexual revolution.

D. W. WINNICOTT ONCE MUSED on the importance of breathing in exploring the question of birth trauma. He says he learned about it from a patient who explained:

> At the beginning the individual is like a bubble. If the pressure from outside actively adapts to the pressure within, then the bubble is the significant thing, that is to say the infant's self. If, however, the environmental pressure is greater or less than the pressure within the bubble, then it is not the bubble that is important but the environment. The bubble adapts to the outside pressure.

This description is the precise state of things when it comes to birth for Winnicott—the character of the bursting of the bubble, the importance of the inability to distinguish a baby from its environment.

For Winnicott, the first impingements from outside the infant's self, in the environment, happen in the birth experience, from the contractions of the uterus and the crushing of the head, to the changes in temperature, light, noise, leading to the necessity for air. Up until then, the fetus was able to go along just being, developing, growing (of course Bion disagrees, and yet sees birth as an extreme interruption). Birth suddenly made the fetus into a "reactor" as their "life world" is withdrawn from them.

In a fascinating and sweeping statement in "Birth Memories, Birth Trauma, and Anxiety," Winnicott declares that birth trauma can lead to what he calls "congenital (not inherited) paranoia." He goes on to outline a series of paranoid, "hypochondriacal" reactions in adult life that are related to the constriction of breath at birth. One of which is to try to derive pleasure from constriction, to bring it under one's control rather than suffer it passively. He says we see this often enough in perversions that involve the constriction of breath. Autoerotic asphyxiation is the most obvious correlate, but there are other forms of what is now called "erotic breath play." Perusing pictures of AADs (autoerotic asphyxiation deaths) collected to help determine the difference between suicide and these accidental scenarios, one couldn't help but see the various contortions of their bodily positions—naked, often face down, tied by multiple ropes, hands, arms, and legs rendered useless—as some mighty struggle to be born.

I will always think of Michael Hutchence from INXS, who either died by suicide, belt around neck attached to his hotel room door, or accidental autoerotic asphyxiation, as his wife claims. After the death of Hutchence, his wife Paula Yates publicly fell apart and eventually died by heroin overdose. Her death was followed by the death of her daughter Peaches, also of a heroin overdose. This family tragedy seems to outline this confused admixture of pleasure and pain and asphyxiation that radiates outwards and through generations.

While there is certainly a long-held belief that one can intensify pleasure from suffocation, there isn't really any proof that it is so. Doctors in the seventeenth century thought asphyxiation could help with erectile dysfunction, and there were a variety of methods applied that led—in the eighteenth century—to a fad for "hanging men's clubs." The whole palaver was said to spring from seeing that men who were hanged sometimes got erections—"angel lust." Beckett gives this absurdity a nod in *Waiting for Godot*:

> VLADIMIR: What do we do now?
>
> ESTRAGON: Wait.
>
> VLADIMIR: Yes, but while waiting.
>
> ESTRAGON: What about hanging ourselves?
>
> VLADIMIR: Hmm. It'd give us an erection.
>
> ESTRAGON: (*highly excited*) An erection! [. . .] Let's hang ourselves immediately!

There seems to be some phantasmatic confusion between erection and breath holding, as if you can hold on to it and fill it up with your inhalation—a fantasy that helps avoid both dying and detumescing like a saggy balloon. Physiologically wrong (as most fantasies are), the occasional "angel lust" happens when blood is cut off at the neck and pools downwards. It has nothing to do with cutting off breath.

Winnicott being Winnicott notes something sweet in all this madness. He says in the erotic asphyxiation of a lover: "active suffocating can be a perverted kindness, the active person feeling that the passive one must be longing to be suffocated." Wryly, he suggests that there is something of this confusion in all healthy passionate encounters. A recent article in *The New York Times* by Peggy Orenstein spoke to an increase in teenagers strangling each other during sex, shifting from 25 to 40 percent in reports by students on college campuses. Both the stranglers and the strangled thought the other wanted it, and neither seemed to know where this attribution of desire came from.

I had a patient who played with autoerotic asphyxiation, sometimes adding drugs that increased the feeling of asphyxiation, from poppers to nitrous oxide to smoking crystal methamphetamine. He also played with contracting HIV during these sexual encounters, refusing to use protection. I had to be incredibly strong not to give into an anxiety that would not help him in the least, especially if it betrayed any feeling of reprobation that he played with provoking.

I have never altered my voice as much as I have with this patient; I brought it down to a near whisper. His prior psychoanalyst told me that he was unable to create a space free from aggressive competition and rivalry. I noted this without knowing what to do with it. This whispering, mind you, wasn't immediate. It developed in the intimacy between us. I didn't entirely know why, but it felt like a minimal figure of breath, an attempt to find an unconstrained breath that wasn't tied to death, asphyxiation, or rage. Bringing us closer to Eros. As I whispered, he started to say he liked coming to see me—that I was so cute, maybe even kind. The terse atmosphere lifted.

My patient struggled with institutionalized sexual abuse at the hands of teachers and professors that followed him well into adulthood and postdoctoral studies. As a gay man he wondered if he had deserved this or wanted it. He wondered whether he was intellectually equal to them—did they see something in him, or did they tolerate him to be close to his physical beauty? He assured me that he was especially beautiful as a young man. (I don't doubt it!) All this confusion running between sexuality and intellectual life, masters and disciples, made it difficult to pursue his own interests. He felt abandoned by these mentors while enraged that they preyed on him, filled with a nostalgic longing for the moment he was the coveted boy full of promise. He tried to impress me with the stories of who he had met in his previous life.

My patient's other symptom was hoarding. He lived among ceiling-high piles of old newspapers, books, items collected on the street, and the food and waste of his one companion—a feral cat he rescued who attacked him every time he came home. The collection of books redoubled as artefacts that recounted his history (he *loved* history books). His space was as asphyxiating as his sexual practices. I heard Michelet, in the archive reconstructing French history—"I breathed in their dust."

Much of his hoarding concerned objects kept for some future, while tied to a past of cherished thinking and writing that he hated seeing being tossed away as so much garbage. "I wanted to be rescued," he exclaimed one day. "I thought that was what they were doing . . . The gay

French intelligentsia was only concerned with itself." The image of himself as suffocated, watched, used, dropped, disposed, emerged as we moved back and forth between his world at home and these sexual encounters.

These images were so painfully solidified, especially his feeling of envy for their power and fame. Digging around in these memories impelled him to seek the point of fainting or losing consciousness even more vehemently. Would he master this severe master? Would he come out the other end without dying? I didn't know. I had to tolerate my distress not knowing what would happen to him to continue working with him. I had to keep whispering to him about his life and desires, like Catherine Vanier to those babies who would not breathe on their own. Her premature infants tested the limits of others too, while asking to be treated as part of the living world. I had to continue to hope that he would shatter these impossible imagoes.

After an anxiety-filled week of waiting for yet another HIV test, and a moment of losing consciousness that finally, truly terrified him, something began to shift. He described the moment of blacking out, his legs coming out from under him. What if he died? He seemingly had to come as close as possible to suffocating before leaving this sexual practice behind. I felt his depression shake as he spoke the first words of concern for himself. The brutality of his sex partners emerged. They left him to handle himself after falling while knowing he was older and more fragile. Why entrust them with your life? This wasn't a space for intimacy or rescue. Neither was French academic life. We moved on to the task of finally cleaning his house.

I REMEMBER A Lacanian analyst from Paris coming to visit my office and looking at the tissue box next to the patient chair with disdain. "Is this for you?" he asked. "We don't even have bathrooms for our patients. They can put their shit somewhere else." "That wouldn't fly in America," I retorted, though I did note to myself that not having tissues there to patronize or prompt patients might be a good thing.

The idea of crying in therapy has become cliché, making it less likely for this cry to command a return to life. Hence the momentary fad for primal scream therapy, as if upping the ante is the way to solve this problem. Now there are just too many scenes on TV and in movies, too many memes on having a "good cry." And yet it still manages to happen from time to time. For Winnicott the cry can have "an expulsive function with clear aim, to live one's own way and not reactively." Neither self-pity nor catharsis, the cry can be the complete exhalation that is a hearty assertion of self; it can finally establish a point of difference between you now, in this session, and all that has happened.

In an utterly embarrassing memory of an early session during my first psychoanalysis at the age of seventeen, the analyst asked me what made me cry. I shrugged my shoulders, looked at him blankly, and said, "Dead puppies?" I can't remember what the sentiment was in this reply. I just remember saying it. Was I as snide as I seem? Actually, this is a more telling answer than one might imagine, but it would take a while to get to *that* part of the story.

Sometime later I discovered a picture I had drawn of myself with huge tears coming out of my eyes that I sent to my mother in the Philippines after I accidentally killed my first pet hamster. "I HAT MAYSEF" was scrawled across the top. Many more hamsters would come to die after this one, drowning in the pool after they escaped their cage. There were also dogs that died, as well as parrots, tropical fish, lizards. A real Florida story of too many exotic pets.

Speaking about these pets in psychoanalysis, a horror at the absurdity of allowing so much death crept up. Why were these animals collected and allowed to die? What was this murderous possessiveness and neglect? Somewhere I must have identified with these discarded animals. While I had been deadpan when I gave that absurd answer to the question of my sadness, when I was finally drawn back to the memory of my first hamster, I found the emotion that hadn't been there.

I saw myself chasing the hamster, how it got squished between the door and the door frame. I felt like I broke it. I remember locking myself in my room, hysterically sobbing. The aliveness of this scene felt distinct from the other memories. I felt lost with respect to these other pets, not merely emotionally but where I was. I couldn't tell where the violence started or ended, whether I felt guilty, or who I might be angry at or want to hold accountable. This is what Winnicott means by a loss of identity. The psychoanalyst in me now hears the wish in the written mistake—the "may" in "maysef" seeking permission to just be.

Thinking of that sick parade of dying pets, I recall how Winnicott turns to the somatic vicissitudes of birth trauma one last time. He notes that the feeling of having difficulty breathing can become the idea that one is lacking something. One imagines this lack could be relieved if breathing could be freed. One can get very attached to an idea like this, says Winnicott, recalling a six-year-old who learned about oxygen and seized on it immediately by constantly complaining that she needed more of it. One could read behind this scene all kinds of phantasmatic ideas about freedom.

Patients, Winnicott notes, like to hold on to the feeling of deprivation, to count their losses, as if doing so might be a protection against further hurt. However, if, through analysis, the patient can remember this primary collection of "lacks" that they carry around with them, they might finally be able to forget them. But, he notes, they must do this one by one. One memory of persecution at a time. One pet at a time. Breath by breath. In this slow, incremental way, psychoanalysis attempts to allow the person to return to just being, and breathing.

IN 1924, THE MYSTIC teacher George Gurdjieff wrote: "all Europe has gone mad about breathing exercises. For four or five years I have made money by treating people who had ruined their breathing by such methods." A century on, seemingly everyone engages in a daily dose of therapeutic breathing. I even sense the remnants of primal scream therapy—so popular in the 1970s—in the new trend for heavy Wim Hof breathing in a dark room with thumping techno music accompanied by shouts of encouragement from a "guide." This current rush towards breath feels backlit by an image of freedom that obscures the violent takeover of the air that defines our century.

When I think of the need to breathe, more so than any spiritual meditative breathing exercises, I think of dancing. Yoga is close to a form of dance for me, and I really do like dancing late into the night, packed together with others. I often feel a kind of itching if I haven't done it in a long time. Freud, it must be said, was never a fan of music or dance or even alcohol. Despite his early experiments with cocaine, he opted for something closer to sobriety, looking for what can appear after the outlines of disappointment have been wrestled with and limitations accepted.

There isn't a lot of the ecstatic or liberatory in Freud's day-to-day universe—hence his total rejection of catharsis. Yeah, yeah, I know, Freud's image of ordinary or common human unhappiness feels rather uninspiring. Who wants to be common? But isn't that the point? To abandon this selfish exceptionalism? We escape from our responsibility to life among others, carefully tucked away in our air-conditioned domestic spaces while fires and floods are *their* misfortune.

Perhaps the quotidian could be seen as an achievement, and finding one's way along the long, tortured pathway towards it celebrated. While breathing therapies go down much better among the public than psychoanalysis (which people accuse of being elitist and impractical), in fact Freud felt that many psychological and spiritual promises were themselves grandiose. He seemed to intuit that the pursuit of breathing freely was vulnerable to forms of paranoid control in the attempt to achieve victorious freedom. He saw in it a desire to fortify the self, not take the self apart in the direction of others.

Freud pointed out in *Civilization and Its Discontents* that man was probably freest before civilized life, but then this freedom had little value since one was "scarcely in a position to defend it." The desire for freedom—inherent in the quest for easy breathing—could be directed at some form of injustice wrought by the particulars of an historical epoch thus bringing about something new in the name of greater equality. But the desire for freedom might also be a wish that we remain untamed by civilization, hostile to civilization altogether.

This form of hostility, for Freud, is no small part of the destructiveness of humans. Why? Nature and other humans engender unhappiness, conflict, and limitation, and it will always be so. This hostility betrays a refusal to recognize the renunciations that civilization imposes on all of us—a refusal that Freud felt was a wholesale rejection of reality. Living with others, caring for others, always means less, not more, freedom.

In Freud's terms, the dream of finding the right breath—whether that's in the form of cloud busting or the various New Age spiritualities—can culminate in an image of blissful oneness, whether that's oneness within the self, with the world, or with others. While I have no end of praise for breathing one's way into and through discomfort, breathing like an ecstatic song following the deep rhythms of a singular body, breathing to return to what it means to carry on being, especially being intimately with others, we must nevertheless be careful. Breathing as oceanic oneness might be a force of denial—denial of our separateness *and* of our continued conflictual interconnectedness. This message seems to be my cautionary tale as a fellow pulmonaut.

Pursuing breathing in the twenty-first century landscape has meant questioning the idea of a psychoanalytic cure and reckoning with the violence and loss that is increasingly being marked by pockets of asphyxiation that appear throughout history. The personal pursuit of liberation so often works in tandem with violent systemic forces that are backed by the same reasoning. This is the breathing hole I keep finding myself in, bouncing between the psychological history of free or constricted breath and a history of violence against breath and air.

Traversing so many wild speculations on the implications of breathing, a whole history of forgotten breath therapies are hidden behind the advent of so many new ones. Looking at therapies designed to counteract or reconstruct an experience of birth, I started to feel the presence of another pocket of asphyxiation—one quite intimate to myself. As someone who has given birth twice, my "labor" seemed altogether missing from the scene. As Irigaray claims, the forgetting of air is deeply tied to an erasure of the feminine.

Doesn't the experience of radical separation belong to those who give birth as well as, if not more than, the child? Why was the one who has difficulty with birth always the child and never the mother? How many men do we need professing the philosophical importance of time in utero, the traumatic sequelae of birth, and the consequences of the wish to return to this intimate space? Did anyone ask us if we wanted them to return? If the womb is such a desired and fearful place, does anyone want to think about what it means to be this place of such extreme ambivalence? For all this murmur about the trouble with being born, what about those who bore?

As for pain, in response to the question of where all the beautiful hysterics of Freud's time have gone, many psychoanalysts have said that we should seek them in pain clinics. Pain unites the body, voice, and breathing and gives painful psychic feelings a place, a locale. As history would have it, the question of women's pain is something no one wants to hear about. Also, their grief. Allow me this last detour on breath and death.

In *The Gender of Sound*, Anne Carson asks why female sounds are so bad to hear. In Hellenistic and Roman times, physicians often recommended vocal exercises to cure all manner of physical and psychological difficulties—"the practice of declamation would relieve congestion in the head and correct the damage that men habitually do to themselves in daily life by using the voice for highpitched sounds, loud shouting or aimless conversation." These therapies were not recommended to women or eunuchs, who didn't possess the right kind of flesh for the low vocal pitch that was seen as curative. Women's voices— deafening shrieking, groaning, and moaning—are seen as seeking to seduce, distract, torture, disorder, stupefy, and drive the other insane. Sometimes all three, as in the case of the sirens who run men's ships aground.

In Ancient Greece, women had a particular shriek during rituals named the *ololyga*—an onomatopoeic derivation that travels in the company of *eleleu* and *alalazo* in Indo-European languages. What I find so beautiful is that these words for women's sounds do not signify anything except their quality as sound-making. The word is the sound. And the sound that they make is a cry of intense pleasure or intense pain—or, most likely, the world-order-defying combination of the two.

This eruption of sounds—so much more affect-laden than signifying—transformed during COVID from the banging of pans, clapping, and shouting of Manhattan's one and a half million residents in lockdown for the 6:00 pm shift change of hospital workers, to the sounds of protests in the streets, the repeated cry of "Black lives matter," and the listing of names of those shot dead by police. The constant sounds of ambulances were drowned out by the sounds of police cars in pursuit of the protesters. What was an intense feeling of pleasure and gratitude, almost collective awe, became a feeling of shock and pain—the sounds of unending grief. We passed from a feeling of world-making to a feeling of world-destruction in a time frame that was only a breath.

These sounds/words follow what was not permitted in civic spaces in Ancient Greek society. There, the sounds of women should not be heard by men, especially men going into battle—as if it would stop them from murdering effectively. Verbal continence is praised and seen as a sign of masculine virtue. My son likes to quote Tony Soprano when Dr. Melfi tries to make him talk about his panic attack when his beloved ducks flew away—"Whatever happened to Gary Cooper, the strong silent type? That was an American. He wasn't in touch with his feelings. He just did what he had to do." "Those are strong feelings," Dr. Melfi replies sardonically.

Carson goes on to note that in Ancient Greek, mouth and vagina are indicated by the word *stoma*, with an adverb to denote upper or lower mouth. "Both the vocal and the genital mouth are connected to the body by a neck (*auchen* in Greek, *cervix* in Latin). Both mouths provide access to a hollow cavity which is guarded by lips that are best kept closed." And many physicians would treat both mouths if there was a problem with only one. In general, women's mouths are best kept locked or closed, though the task is a difficult one as women tend to be condemned as leaky vessels like the jar of the Danaids. They can't help but let out these unspeakable things.

According to the classicist Nicole Loraux, mothers in mourning are the most dangerous threat to the civic order. The sound of mothers' grief could undo an entire nation, especially in times of war. Mourning tends to be spoken of as feminine in its essence, an idea that seems to back the limited roles allotted to women in general in Greek antiquity. In fact, the link between mourning and femininity is so strong it isn't clear which one is the primary danger—whether mourning must be contained, and women excessively mourn, or whether women must be contained, and so too their apparent excess in the face of grief.

The rules restricting women's mourning make clear that it isn't the grief as such, but being witness to her pain, especially the sound of it, that is the danger. "And the *páthos* is piercing: *álaston odúromai*, 'I grieve without forgetting,' says Eumaios to Odysseus. Or rather: Never do I forget to grieve, I cannot stop grieving . . . *álaston* expresses the atemporal duration, immobilized in a negative will, immortalizing the past in the present." Loraux draws out the question of mourning around amnesty and amnesia, their enticing alliteration, as if born one from the other. If women refuse to either forget or stop grieving, the city and its institutions must require oblivion for the sake of amnesty. Forgiveness is forgetting. You see, like Freud, the ancient Greeks understood that some repression is necessary. But then, it always goes too far, attacking the most helpless.

Political authority in Greece set up a ban on remembering certain misfortunes several times. Women's mourning was then confined in time and place, and finally erased. Public funeral oration in Athens excluded forms of lamentation and complaint in favor of praise or eulogy. This would be men's speech. The nation was increasingly seen as built on the sacrifice and devotion of its citizens—the demos—and lamentation was putting one's personal lot ahead of the social. An excess ascribed to feminine flesh must be domesticated to prevent self-annihilation, civil war, or, worse, murder. Epidemics of suicide by young women, they said, must be prevented.

When the social body is sick, the nation is in danger of civil war. This was seen by the Greeks as a body that failed to breathe, whose borders were closing and causing asphyxiation. Loraux paints a portrait of the Greek ideal as a body that must be open—a logic that makes the good death that of the warrior who is cut in battle, or the animal who is slaughtered as a gift to the gods. This shedding of blood is of value to society.

Women are the ones who often die by the noose—necklace of death—off-stage and in silence. Their bodies were always seen as in danger of asphyxiation, especially *hysteria*, which means a wondering womb that strangles from the inside. "This can probably be seen as a way of denying the "simple" evidence that women's bodies are inherently open—slit." Also, that women bleed naturally for the purpose of reproducing citizens.

Despite this strange co-opting of the female body by the world-making of men, women are nonetheless equated with the force of *stasis*—stoppage of time, by negative internal will, excessive mourning, and extreme emotional pain. "Hanging is the lot of desperate people who have lost all time," writes Loraux. There are only two incidents of hanging in Thucydides and they are blamed on the horror of *stasis*. Stasis means being unable to tell the difference between inside and outside, suicide and murder, brother and enemy, household and city—civil war abolishing all limits. "Surrounded by their democratic opponents, caught in a trap, reduced to absolute disarray, twice the oligarchs of Corcyra are forced into suicide, either hanging themselves from the trees or strangling themselves with the straps of their bedding or strips torn from their clothing."

These are the fascinating, gendered origins of the Western world—cradle of democracy. Nothing is more ignoble or depraved than strangulation, which is the sign of civil strife—a strange feminine infection of a nation. Following the physicians of their time: the body must release its heat, it must breathe. A continual exchange between inside and outside is necessary for health. The hanged person is a body that is forever closed, whose soul doesn't have the means to escape. These ignoble deaths will remain unspoken by the polis.

Women are caught in this image of civic life as being too open and too closed, too labile and too hardened into position, dragging them into the shadows and silence. Too airy or too suffocating. Indeed, women are always a shadow pandemic in moments that threaten an unravelling of a male-dominated political fabric. Women are our containers of grief, our silent breathless screamers. I feel like I've been screaming in this chapter. Also, trying to mourn.

Is it so surprising that a large part of this story of breath is about the troubled relationship between a father of the air and a daughter on the ground? The story ends, or rather begins again, with his dying and my having a daughter myself. Her birth spelled his end, showing him all that he would not live to see—the man who saw everything flying the friendly skies. I saw this on his face when he first looked at her. Women, and other minorities, have, for a long time now, had to carry this projection of death. It makes us one with the air, which (of all the elements) has the closest affinity with nothingness.

LAST WORDS

PLANTS EXIST IN between: rooted to the earth but reaching into the air. Plants are essential to the creation of the air we swim in. They recycle our waste products and produce oxygen. Photosynthesis generates carbohydrates and oxygen in equal measures and fuels our planet. For philosopher Emanuele Coccia, plants are essential to an ontology of mixture; they are cosmic engineers that mix the elements: water, the sun, the air, and the nutrients of the earth. Unfortunately, their work is undervalued, and over the last year alone 988 million acres of land were destroyed by fires around the world, emitting 6.5 billion tons of carbon dioxide.

There is nothing individual about a plant, not even the tree, which is so often used as a poetic symbol for man. The tree must be unnaturally isolated for this poetic trope to work, the lone oak battered by the wind. Plant life, for the main, is pure multiplication, repetition, as much horizontal as vertical in structure. Dense and unmanicured. It relies on so many elements for its continued life—symbiotic, parasitic, and atmospheric. Think here of the importance of worms, birds, bees, fungi, feces, mist, reflective surfaces, and other plants.

Anaxagoras, an Ancient Greek philosopher of nature, saw life as mixture such that everything was in everything. In his concern for infinite mixing he must have been thinking of plants, atmosphere, and breathing. For Anaxagoras everything moves through us, and we are open to everything—a vision of radical porousness not dissimilar to Judith Butler's observation, after the pandemic, that we cannot "live without each other, without finding ourselves inside another's pores." While this might sound like fusion, Coccia wants us to think of mixture as bearing in it the distinctions that are the wondrous diversity of life that plants exemplify—to think of mixture not as unification and the dissolving of differences but as endless diversification. "Breath," he writes, "is the art of mixture, what allows each object to mix with the rest of the objects, to immerse itself in them."

What Coccia calls an art of mixture requires recognizing that immersion is our basic state: immersed in the waters of the womb. Immersed in air. Immersion signals a lack of resistance to life, an ease that is prior to even the wish for nutrition that is a first annexation of the exterior. Could we willingly free ourselves, not into the ecstasy of nothingness, nor into our individual selves, but into this immersion where there are no stable frontiers? Could we see space not as ours to occupy but as filled with so many others and so much life to mix with?

Coccia's reflections on immersion are very different from our usual way of relating to the world around us. Somewhere Peter Sloterdijk wonders at what point weather became a drama to be reported, as if bad weather was unnatural. He is pointing to our tendency to anthropomorphize nature, imbue it with morality, as if it were supposed to conform to our desires and is therefore bad when it does not.

"Why should nature be made to serve as a gigantic echo chamber for the moral orders that human's make?" asks Lorraine Daston in *Against Nature*, pointing out that this moralizing view of nature undergirds the way we engage with the debate about climate change, which is often seen as nature's revenge against human crimes. Is it?

Given our immersion in an atmosphere that we share with all life-forms, our destruction of the natural world amounts to suicide. "Ecocide is always also egocide," writes my good friend, the feminist philosopher Elissa Marder, who points out that the term *climate change* misses the mark, "because climate *is* change." "The problem now for humans," she says, "is that that the acceleration of that change has undermined the fundamental possibility of establishing a relation to the external world as that world has become too unstable."

Daston argues that in the extension of divine creation to the idea of universal natural laws, everything, even the irregularities, were to be foreseen in the same manner as God. We live under the shadow of this religious vision today, with nature still mostly viewed in these moral terms. Chaos is seen as unnatural, which is to say ungodly, when really there is much about nature that refuses to conform to our deterministic way of viewing it. Daston names three emotions that speak to the experience of peering into a world of chaos and unpredictability that lies behind the moral order that we have imposed upon nature: horror, terror, and wonder.

In this book so far, we have spent a great deal of time with the first two. But what about wonder? Wonder interrupts the vision of a deterministic universe. It is an anathema to our attempts to create a final order. Wonder is about the limits of human understanding, arising in the face of chance events, unpredictable acts, whatever could be characterized as new. If we are looking for what breaks the chain of our human reasoning, wonder alerts us to its presence. Wonder speaks to a nature that is full of cracks and fissures, random mutations. Nature seen as the hodgepodge montage that it is: barely fitting together, Frankenstein-like.

For Daston, the full diversity of nature can only be seen when we are closest to its chaos, when we transcend the narrowness of our moral view of nature and our habitual ways of viewing the world. I can't help but think here of the basic psychoanalytic principle of inducing regression in patients as a controlled descent towards this place—name it what you will: chaos, nature, the unconscious, the Real, the navel of representations reaching down into a further unknown. As Freud put it once, "With neurotics it is as though we were in a prehistoric landscape—for instance, in the Jurassic. The great saurians are still running about; the horsetails grow as high as palms."

How might we work better with this human refusal of limitation—a specific human will to project itself forever outwards? How can we jam the machine that wants to determine and control a final image of life? Breathing, as we have seen, is the weak natural link in our bodily order. It is a crucial part of evolution in the direction of some of the most gratuitous and disruptive human acts—namely speaking. We don't know where this speaking will go, what it will create, or what it will destroy. This is why "say anything" is the opening act of every psychoanalysis.

Language is as monstrous as chaotic nature, an assemblage of reversals of meaning and mutations of sound. Always in danger of being overwhelmed by nonsense or emotions. The way language mixes the functions of throat, larynx, palate, tongue, teeth, diaphragm, and nasal system is an unruly bodily montage. As the poet Paul Celan said of poetry and of art in general, it is the path towards something truly Other—an *atemwende* or "breathturning."

IN *UNDROWNED: BLACK Feminist Lessons from Marine Mammals*, Alexis Pauline Gumbs attempts to create "an artefact and tool for breath retaining and interspecies ancestral listening." What if we could breathe like whales and sing underwater, or transform like river dolphins, which in muddy conditions come to rely less on sight and more on sound? Can't we see in these marine mammals not individual survivors but surprising and extreme communal adaptations to a changing environment?

Whales and dolphins are the enviable ones who, in fulfilment of Ferenczi's desire, returned to the nourishing waters. Re-adapting to the water as mammals, their nasal passages transformed into blowholes for breathing, completely defunctionalizing their noses for smelling purposes—lucky, since the smell of the air coming from their blowholes is rank. Fishermen's tales claimed that getting near a whale's blowhole could melt your skin off.

Cetacean breathing is completely under conscious control. In *Moby-Dick*, Melville was awed by the whale's relative freedom from what, for man, is a perpetual obligation:

> In man, breathing is incessantly going on—one breath serving for two or three pulsations only; so that whatever other business he has to attend to, waking or sleeping, breathe he must, or die he will. But the Sperm Whale breathes only about one seventh or Sunday of his time.

The whale's ability to hold its breath for long periods of time and descend into the deep is enviable. Man would do it if he could. He would plug up his nostrils if all the blood in his body could be aerated with one breath. "For not by hook or by net could this vast Leviathan be caught, when sailing a thousand fathoms beneath the sunlight." However, dependence on air is a mammalian plight. It marks the whale as a creature that once walked on land. As Melville explains, it is the surfacing of the whale to breathe that makes it possible to hunt; its need to breathe creates the fatal exposure.

Still, there are advantages to this dependence on air. There is no music without breath. The melancholy music of whales is thought to be for the purposes of mating. Whale song was first recorded in the 1960s by a naval engineer who had been listening for Soviet submarines. Humans have since dreamed of finally being able to decode the language of whales, a dream recently reawakened by the cetacean translation initiative (CETI), an AI project to the tune of $33 million.

Cetacean language is the most nasal of all languages. Forcing air through a series of sacs and flaps, these vocalizations—whistling, clicking, screeching, and pulsed sounds—are both hierarchically organized and yet open in the same way our human symbolic system is. Some dolphin and whale linguists point out a three-dimensionality to their language since they can manipulate the direction of sounds as a means of echolocation. We humans are less precise about our bodies in space when speaking. In fact, breathing is often a means of restoring a feeling of location and orientation that gets away from us when speaking, especially the speaking we do in our minds that we call thinking.

Spoken language is made of breathed sounds. The world of the word is as tied to the production of sound—made from a column of air suspended inside our bodies—as it is to the abstract ideas we call *thought*, or the world out there that words are meant to convey. The word *convey* in Ancient Greek is *metaphor*. I always loved seeing the luggage trucks with "metaphor" written on them when landing on the tarmac in Greece. I think one of the greatest wonders of life is how language lives, how it moves, how it engenders new realities.

While Freud was working out his theory of the psyche, the Swiss linguist Ferdinand de Saussure revolutionized the science of linguistics. His work would inspire not only philosophy and modern literature but also neuroscience, psychology, and anthropology. He devised a way of studying the structure of language that wasn't a reduction of language to grammar or comparative philology. His new science was able to circumscribe an essence in language.

From the vantage point of linguistics, humans are separated from other forms of life because we are bound as a community through an exteriorization of ourselves in the transmissible traces of language. Of course we are approaching the word *culture*. It's a word that annoys me. The word *culture* evokes the image of a stroll through a museum, as if it could simply be surveyed. Really, we are up to our ears in it, and yet, ubiquitous to the point of invisibility, we can barely see it. The problem is that *culture* doesn't capture how language marks us on a cellular level, the bidirectional infection of language and life. "Language is a virus," said William Burroughs. Once it gets in our cells, it never leaves.

"To speak is to feed," writes Michel Serres, who describes the speaking human as a sort of parasite, feeding on air. Isn't the truth of a parasite both its voracity and its fragile dependence? An inability to ever be weaned— to "grow up," become independent? From the perspective of language, we absorb everything from it, as if into our very skin. We are as defenseless to language as we are to air. We repeat what we incorporate from it and meagerly return the gift.

Before speaking, a baby's babbling makes all the possible sounds that compose human languages. If one does begin to speak, language is reduced to the differential sounds of the child's early caregivers. Our first words contain the entirety of language as the subtraction of the languages that will never be spoken unless learned later, built upon one's mother tongues.

Still, don't get too fascinated by the child who works to apprehend language, Lacan warns. It obscures the fact that the "symbol is already there, that it is enormous and englobes him from all sides." Language is not our tool, rather we are the tool of language, as a colleague once put it to me. Lacan makes a remarkable link between the intrusion of breath and the intrusion of language, both of which we must learn to swim in:

> By emerging into this world where he must breathe, first and foremost he is literally choked, suffocated. This is what has been called trauma—there is no other—the trauma of birth which is not separation from the mother but the inhalation, into oneself, of a fundamentally Other environment.

For Lacan, there is something foreign and traumatic about the unending presence of air and of language. The human wish to dominate belies the fact that we are subordinated by the perpetual obligations of language and breathing. The rejection of this dependency makes us prone to all manner of violent resistance, rejection, denial, and forgetting. If ever there was an image of a decentered center in human life it would be breath: "Without ever being able to habituate ourselves to the air, or fully to inhabit it, we have nevertheless brought the air under occupation. The air has been finitized, unothered," writes Steven Connor in *The Matter of Air*.

Do we really want to be part of the community of breathing, speaking humans, bound to one another? Can we accept what in breath, language, and nature will outlive us? Perhaps air and language should sit alongside Darwinian evolution, the Copernican doctrine of heliocentrism, and the unconscious on Freud's famous list of "blows" against human narcissism. Psychoanalysis tries to locate what gets disavowed or "forgotten"; most notable are sexuality, language, and air—these infinite fields where we cannot centralize any sense of self.

Lacan would extend forgetting to the fact of speaking. Our attention is always drawn to the feeling that we "think." We lose touch with the desirous reach towards others through speaking. We also actively forget about the actual act of speaking, which requires the production of sound, the movement of lips, the placement of the tongue, the manipulation of air. Speaking must happen in real time, in an intense confrontation and interpersonal exchange. There is the bodily and energetic dimension of trying to say, having to say.

I think we have real confusion about the conversations played out in our mind and those we have in real life. IRL, to use today's neologism, which is extra funny since its usually written, not spoken. The ubiquity of texting instead of speaking seems to have led many of us to be terrified of hearing voices. My son brags that he hasn't listened to a voicemail in four years. I'm amused (and bewildered) by how easily I've forgotten how to talk to friends on the phone, something I did incessantly as a teenager.

We also forget the speech we have heard and taken in—as speech (as Hamlet said, "I eat the air, promise-crammed"). We lose hold of the speech that has impacted us, lives in us, toys with us, parasites our memories and thoughts. We forget that it is something said, composed of sound and air. We forget that when we felt bad there was someone there who said something to us, about us; we are spoken to and sometimes obliterated by what others say to us. Sometimes we try to reply in our minds, craft an airtight defense.

Psychoanalysis tries to detoxify this speech—open it back up, allow some ambiguity to aerate it, some room for the other. I was always so moved by the idea of the end of a psychoanalysis as a radical moment in which the speaking partners stop; they stop speaking. To make the end a real end, there must be a feeling that the stopping means that the relationship cannot just resume at any moment. It's a rare opportunity that signals an experience of an end that isn't death. What was held and carried by speech returns to memory, sound.

Psychoanalysis helps navigate the Scylla and Charybdis of language and sexuality. In this world of competing therapies (where psychoanalysis is especially vulnerable to the charge of intellectualization and the navel-gazing activity of talking about oneself), I think we must see this bodily breathing dimension. Breath unites the two hazards of sexuality and language. And so, to the forgetting of sexuality and language, I add breathing. In *Le Stade du respir*, the philosopher Jean-Louis Tristani argued that before desire, there is breathing, and that remembering breathing is the path to revitalizing desire. In breath we find the knot of our forgettings. The hope is that breathing is remembering.

IN 2018 THE state of Alabama executed Kenneth Smith using nitrogen hypoxia, forcing him to inhale pure nitrogen through a mask until he suffocated. Despite state claims about this being a painless way to die, Smith thrashed around for two to four minutes while attempting to hold his breath, and this was followed by five minutes of heavy breathing, involuntary movements, and gasping. In the words of a human rights chief speaking at the United Nations: "This novel and untested method of suffocation by nitrogen gas may amount to torture, or a cruel, inhuman and degrading treatment."

America was the first to develop gas chamber technology, well before World War II, and it is now the first to use nitrogen as a new technology for execution by gas asphyxiation. While these may seem like extreme examples, this is a particularly visceral version of a degradation that increasing numbers of us are facing. Each year as many as nine million people die due to inhaling toxic gases and airborne particulate matter from fossil fuels—just one of the lethal impacts of widespread environmental pollution, which also extends to black lung in miners, poisoned water, and fatalities in climate-driven catastrophes such as wildfires, extreme heat, and floods. Breathing toxic air has countless more mundane effects: a recent *New Yorker* article titled "The Hazy Days of Summer" notes that, as air quality decreases, test scores go down, crime goes up, umpires make bad calls, and investors make more mistakes.

This is an example of what Lauren Berlant called "slow death"—defined as the "physical wearing out" of a segment of a population, in which life-building and life-demolishing are utterly confused. Berlant focuses on obesity, the way it has impacted the minoritized poor. Obesity is so readily seen as a personal failure of will, while the availability of healthy food is unevenly distributed, fast-food industries have been allowed to proliferate, and generations of impoverished families are bound by the pleasure of eating in a life that is offered few spaces for relief. Berlant sees obesity and slow death through the prism of what, in their book of the same name, they term "cruel optimism": our attachments to problematic objects that we fail to detach from because they feel necessary for living.

Cruel Optimism was published in 2011; from our post-pandemic vantage it is easier to see how, in the physical wearing out of marginalized populations, different forms of harm overlap: obesity, disproportionately affecting Black people, increased the rates of death in the pandemic. Slow death and toxic air point to the atmospheric nature of contemporary harms: they represent the complicated and entangled desires that undermine fantasies of the sovereign self. We blame individuals instead of understanding that this is a problem orchestrated by the management of life and profit.

Berlant sees in these symptoms a fight against the demand to be productive and live up to one's potential, as if this were under our individual control. The fantasy of sovereignty, writes Berlant, takes "a militaristic and melodramatic view of individual agency by casting the human as most fully itself when assuming the spectacular posture of performative action," when in reality, our choices are constrained by living in atmospheres that, quite literally, make us overwhelmed, dissipated, worn-out, confused, stressed, and anxious. We live a series of cruel contradictions and inequalities. What action should come from that?

It is hard to separate ourselves from what is already not working. Our attachments are who we are. Anne Carson makes the fascinating observation that in one of St Paul's letters in the New Testament, he describes people wandering in the wilderness eating "pneumatic" bread and drinking from a "pneumatic" rock that followed them. "Can either bread or rock be made of breath? Anyway, who can drink from a rock?" she asks. She also wonders why they couldn't come up with a better, or less awkward, image of God's care. But perhaps it's not such a bad image of care, after all, given our painful entanglements. Breath as a rock that doubles as food speaks to our messy attachments— what sustains us also weighs us down.

In Ferenczi's *Clinical Diary* he details the wish certain patients have that pieces of themselves that feel dead could be revived by an extraordinary love—the epitome of optimism. We want to be whole again. Such happiness will not be had, notes Ferenczi—in fact it perpetuates unhappiness by becoming a means of continued holding out. Still, even if optimism is often cruel, for Ferenczi things are not quite hopeless. Written in 1932, Ferenczi's observations feel as contemporary as Berlant's:

> It is my hope that with tremendous patience and self-sacrifice on our part, after hundreds of instances of enormous forbearance, sympathy, the renunciation of every authoritarian impulse, even acceptance of lessons or help from the patient, it will be possible to make the patient renounce that colossal wish fulfillment and make do with what offers itself.

Of course, this work is slow—beset by resistance.

At one point in his diaries, Ferenczi declares that he finally found the root of his own breath-holding demand for love: the tragic moment in childhood when his mother declared to him, "You are my murderer." How much kindness can we extend to both mother and child in this moment of trying to willingly give up an immersion in one another? Ferenczi asks himself. Both are looking for the place where they can be exposed to a harsher reality. Separation is slow birth, not slow death, though sometimes it feels difficult to disentangle these. "To mix without fusion means to share the same breath," says Coccia. We burst the bubble again—this explosive exposure to the impersonal space between us.

I feel the hard edges of my breath. I hear your breathing.

TWENTY-FIRST-CENTURY MINDFULNESS preaches that all we need to do is breathe, as if nine billion individuals breathing properly will make social problems evaporate. Is all breath work optimistic about what it achieves? How much should we worry about the cruelty of this optimism, especially as air itself becomes a privilege—an increasingly scarce commodity that is unequally distributed?

In *Blackpentecostal Breath: The Aesthetics of Possibility*, Ashon T. Crawley uncovers a history of collective breathing that works as a counterforce to the toxic histories of violence and oppression. Crawley searches Black Christianity for what he calls "fugitive" breath practices: from the powerful enunciation of song to whooping, crying, and speaking in tongues.

Blackpentecostal breath is a performance of displacement where breath is the medium of this movement. Breath happens in flesh and with a voice that is otherwise riveted to its place by racism, slavery, and segregation.

"The fugitive enacts by enunciative force, by desire, by air, by breath, by breathing. Breath and breathing of air, in other words, not only make possible but sustain such movement." Within an atmosphere of anti-Black racism, Blackpentecostal practice speaks to an encircling of a power to move, disrupt, emerge, intensify, that is unique as an embodied and wholly social spirituality:

> Blackpentecostalism privileges interruption, eruption, of air, of breath and breathing as being contingent, on edge, waiting for the perculatory moment of the otherwise breaking into and surprising the service . . . "the pause" . . . "is much more than a break in delivery" but also "an opening in the preacher's consciousness through which the musicality of the Spirit breathes so that the musicality of the sermon resonates with the living truth."

Moments of speech are eclipsed by silence as the community pauses and takes a deep breath together. Then all manner of reverberating sounds begins, the smacking of the tongues, deep panting, whooping, atonal exclamations, the "vibration of nasal passages," until the silence is filled. This fullness asks for the sermon to resume.

Crawley shows how silence is an invitation to excess; excess is invited into the outside. This exterior excess is called praise. This idea runs counter to the anxious or defeated experiences of silence, or excess of silence, that can be felt as menacing. In fact, this opaque silence and joyous excess is what is often projected into Black bodies and punished by the racist. With Blackpentecostal practice the congregation is invited to speak, take, yell, praise, eat, drink, share, or remain silent. Appetite or its refusal are both welcomed.

Julie Beth Napolin, who works in sound studies, has written that the silence of lockdown was the perfect stage for the eruptions of the Black Lives Matter movement. When else could something that has failed to be heard have a chance of being listened to? It was as if the silence of the analytic atmosphere extended into the world. Likewise, in Alex Ross's essay "What Is Noise?" he notes that COVID showed us the sonic havoc we wreak as a species. Driven indoors, machines and cars brought to a halt, birdcalls regained qualities that were only recorded decades ago: singing in lower frequencies, with richer and more complex sounds, or singing more softly. It turns out the birds had been doing something equivalent to screaming and found respite as a species in our relative silence. Thus, there is already a long natural history of knowing what it means to be called by a silence.

Pentecost is also about feasting. "The mystery of trans-substantiation is there; it is clear, luminous, and transparent. Do we ever eat anything else together than the flesh of the word?" asks Michel Serres. So much talking happens while eating. We find in the Pentecostal tradition a foundation for everything that can be done with a mouth, a nose, lungs, a throat, stomach, as part of a shared spirituality. The fleshier the better. This excess is also captured in the practice of speaking in tongues.

Certainly, a form of prayer, a moment of divine inspiration, the question Crawley asks is if this tongue speaking is seen as "on the way to speech." Is it a language of the dispossessed or a xenolalic ability to speak an unknown foreign language, like that of God? Or is it rather a glossolalia, meaning a nonlinguistic eruption, a "speechifying of nothing." He places his bets with the latter, for there is an important right to mean nothing that he sees in Blackpentecostal performance. A worship of what is beyond sense and meaning.

When one is subject to violent erasure, this violence also comes with a demand that one justifies their right to exist, however much this justification will be proven as unjustified. In Blackpentecostal practices there is a space reserved for being willfully obscure and incomprehensible. One doesn't need to justify what is said or move into the realm of making meaning. Speaking in tongues marries desire and speech at this point of nothingness. A kind of opacity or limit to understanding announces itself.

> There are all sorts of gaps and elisions and ruptures of sound, thought, texture, openings and forestallments that go against any such notion of "perfection" and reproduction that could ever be so termed. These are the calling forth, not just the call and response, but call and call, some sort of accretion and accrual, layer upon layer, each word and phrase and scream and breath engaging and revising that which came before it, affecting subsequence.

Glossolalia devalues the individual cognition, which betrayed the Black community. Speaking in tongues refuses to allow understanding to have the final say.

An accretive knowledge the community or congregation doesn't "know" is placed in the foreground. As Caterina Albano acknowledges, there is a direct link between breathing, disappearance, and community in the flux of breathing: "By breathing, we experience both autonomy and commonality as we unconsciously perceive not only our own breathing patterns but also those close to us." Like the breathing machine of the church organ, the congregation is filled with breath, air, wind, in the image of the holy spirit. The congregation produces a cacophony of sound that holds air close.

Maybe we could call this celebration? Or mystical ecstasy? A search for a kind of apex that is equivalent to nothingness, but still togetherness, being-with? Crawley also calls this a moment of compression, condensation, that serves decompression—like deep-breathing exercises that play with retention. One learns that refuge is only for a short duration but not indefinite. Blackpentecostal breathing provides a moment of affirmation against a backdrop of suffering. It takes on the problematic and impossible wish to be relieved once and for all (like Ferenczi's patients).

These are important breath lessons by the dispossessed. Their breath had been taken as someone else's possession. Breathing again can be a disruption of the status quo, a means of resistance, affirmation, and a reassertion of community. It goes against some womb-like imagination of possessing the mother's body, the earth, or even all the others. It also pushes back against any idea of a triumphant autonomous breath. One here is merely passing through. Home is temporary. But a home can be found in the communal staccato breath, the disruption in time, the group uproar!

Everything leans into the momentary. These are the bonds of a community that will not seek violence against an outside enemy, but consolation in a shared refuge. A pattern of flight and escape is something worth following. We all need to get out of what we are in. Community here is one that allows all manner of movements that can be contained by it, no matter how disruptive, fugitive, minimal, or wild.

Finding home in the temporary, in the air that can be shared, stands in stark contrast to a world of appropriation whose violence has been suffered by the Black community and the legacy of slavery. In this time of catastrophe, we will need yet more lessons on breath. As Frantz Fanon wrote: "Open to all the breaths of the world ... My speaking hands tore at the hysterical throat of the world."

We are constantly missing something singular about our relationship to breathing. In "The Universal Right to Breathe," Achille Mbembe writes, "We must answer here and now for our life on Earth *with others* (including viruses) and our shared fate. Such is the injunction this pathogenic period addresses to humankind. It is pathogenic, but also the catabolic period *par excellence*, with the decomposition of bodies, the sorting and expulsion of all sorts of human waste." Mbembe is hoping the time has finally come to acknowledge breathing and understand that freedom cannot happen alone. That it must be universal.

"To come through this constriction would mean that we conceive of breathing beyond its purely biological aspect, and instead as that which we hold in common, that which, by definition, eludes all calculation. By which I mean, the universal right to breath," writes Mbembe. How, he asks, can we discover that all humans and all the earth are one? Yet the "one" pursued by Mbembe is not that which we have seen so far—that of an individual, fusional love, or dissolution into the cosmos, sea, or air. Instead, it is the image of incalculable diverse beings living together in a shared atmosphere.

How could one not agree to a universal right to breathe? And yet, universals, even when defined in the most basic terms of the necessary elements of life, have proven endlessly difficult. Universals also ignore that no one has ever ceded power without a struggle and the struggle will take place on a playing field that is markedly unequal. An image of undifferentiated sameness is too weak for such uneven terrain. Billionaires will buy up the last places on earth with decent air. Recent research showed that already fewer than 10 percent of countries or territories met World Health Organization standards for air quality: a universal right to breathe, but a universal reality of pollution.

A contemporary problem—but it is important to remember that we have long been concerned about pollution. In Ancient Greece, pollution, or *miasma*, referred to moral and physical contamination. The prefix *mia-* denoted defilement or the impairment of a thing's form or integrity: traitors and lawbreakers are *miaroi* in their disregard of the normal constraints necessary for life in society. *Miasma* was seen as contagious and, if not addressed, something that could infect an entire society.

Miasma tended to follow the pathways of the body around birth, sex, and death. Natural events like menstruation, birth, and death could bring *miasma*, as moments when the biological erupts in social life, which must harness or tame its effects. Incest, parricide, touching a corpse—all these things could be a source of contagion requiring catharsis (*katharmos*), or purification and cleansing. There were many such rituals in Ancient Greece, such as when mourners covered themselves in dust—befouling (*miaros*) themselves in sympathy with the deceased. Purification for the Greeks was moral rectification and atonement, rituals that often involved a sacrifice, a form of paying for one's wrongs.

Michel Serres reminds us that the word *pollution* first meant "the desecration of places of worship by excrement, and later the soiling of sheets by ejaculation, usually from masturbation." This is not so far from modernity's sterile hatred of dirt. The difference being, however, that without a concept of the sacred, we now pollute at will. In our total destruction of nature, in a world without communal bonds, what limits are there now?

IN HER ARTICLE "In Defense of Liberal Conspirators," Astra Taylor asks why the response to coronavirus seems to have been an even greater retreat into fantasy and conspiracy. While researching the word *conspiracy*, Taylor finds a surprising history: "It comes from Latin *conspirationem*, 'agreement, union, unanimity,' and *conspirare*, 'to be in agreement; to ally'—or, literally, 'to breathe together.' That," writes Taylor, "is what the more powerful segment of society, the ruling class, have never wanted the rest of us to do—to come together as allies and, god forbid, form unions." Taylor notes how our ability to breathe together has been prevented by a politics of profit that sabotaged progressive collective action. "In a perverse way, many prominent conspiracy theories are a cry for justice and connection, even if they frame the world in a Manichaean binary of good versus evil."

Surely it is no coincidence that the spread of misinformation during the pandemic took place at a moment during which many of us—indeed, the fortunate ones—were trapped in our homes, unable to breathe together. These provided the conditions for the spread of misinformation: COVID was perceived as a hoax, just as climate change and the wildfires in California were not a result of shifting weather patterns due to greenhouse gas emissions but caused by anarchists in cahoots with Black Lives Matter. We had the Capital Hill riots. Believers in QAnon are now members of congress, as, post-COVID, we're increasingly seeing conspiracy narratives that initially arose and spread online appearing IRL.

We are still trying to understand the internet—a paradoxical space of connection in isolation, a breathless reading *of* and writing *to* a world we are not in when we are in this. In our age of information, we are learning how powerless knowledge can be in the face of fantasy, and how easily manipulated it is (and we are). I recently saw a picture reposted of Donald Trump after his indictment for fraud that said: "he is our George Floyd." The reposter wrote: "the internet was a mistake." It's hard to disagree with this sentiment, while we have to admit how well the internet puts on full display the powerful, complex, and entangled desires we have for connection with one another.

Even as the internet might pervert these basic human wishes, the capacity to be in touch with so many inhabitants of the globe, to have access to this much information, is astonishing. Like air and language, we cannot get our heads around it. While we feel like the internet belongs to us, exists in the palm of our hand, it has decentered us from our life. We will have to learn to swim in it, and, like the free diver, to conserve enough energy to return to the surface. Meanwhile, we ought not to forget that fueling our global cloud requires material energy—a draining of resources destructive of the environment.

In Patricia Lockwood's novel *No One Is Talking About This* she evokes the disembodied, airless world of the internet (which she calls "the portal"). "Why did the portal feel so private when you only entered it when you needed to be everywhere?" Lockwood describes that encroaching feeling in which one knows there is still real life to be lived, real things to be done, but where this reality keeps slipping further and further away: "This did not feel like real life, exactly, but nowadays what did?"

The arrival of a baby marks a turning point in Lockwood's novel. The protagonist's sister is pregnant with a baby with a rare genetic disease: the baby, if born, will not live long. Nothing cuts life open like a baby. But a baby that is born and must die almost immediately is the purest wound. In one scene, the protagonist marvels at the sonogram images of the baby "pretending" to breathe, preparing to breathe, practicing breathing—and it fills her with joy. Every person on earth might be given a party for breathing "just like the rest of us." Eventually, the baby is born, and six months and one day after taking her first breath, she takes her last.

All this is set against the backdrop of climate catastrophe: global asphyxiation. "Meanwhile, on the earth of the baby, the climate grew hotter: icebergs melted, the seas rose, permafrost cracked to release prehistory, sections of the Great Barrier Reef blinked out whitely and one by one." Lockwood's description of the baby's dying days leaves a knot in your throat—a knot that should be there every time we hear about the extinction of a species, another step in humanity's incremental slow death. Soon the baby begins to lose her memory of breathing, pausing oddly between breaths as if they might not restart. "If she stops breathing," a nurse tells her mother, "just give her a little pinch on her fingernail." She's simply forgotten to breathe and needs a little reminder. What kind of a pinch do *we* need?

In the final pages, the protagonist's phone is stolen at a concert. She wanted to hear some music, feel bodies pressed against her. Something real. As the phone is lifted out of her pocket, she feels lighter. "Her whole self was on it, if anyone wanted. Someone would try to unlock it later and see the picture of the baby opening her mouth, about to speak, about to say anything." The book ends with the psychoanalyst's invitation to their patient—say anything, say everything. Until reading these lines in *No One Is Talking About This*, I had not heard the invitation to say anything as the mother's desire when greeting her baby's first sound—the patient waiting we do for the moment they will speak. No one *is* talking about this.

Distressed the other evening after being reprimanded, my daughter started shaking and repeatedly saying, "I'm tired, I'm tired," tears falling down her cheeks, which had flushed red. I held her tightly until her body relaxed. I tell her she's tired all the time to explain states of unravel as the time to sleep draws near. But she wasn't tired this time, she was hurt. Maybe she was tired of me, which she doesn't yet know how to say because I say it. Such is our absorption. I marveled at her will to speak through such strong emotions, confront me while seeking comfort.

I know that many feel guilty for having children in an age of climate breakdown and so much social strife. My daughter had to take her first breath in *this* world? This sentiment, the psychoanalyst in me says, is a failure of desire. We must still find the way to want to hear what they will say, to know that there is, while we are here, something left to say, something that hasn't yet been heard. Our eager anticipation means there is still a world.

SCREAMING, CRYING, VOCAL TICS, sneezes, hiccups, laughter, yawning, and orgasm: each of these human behaviors involves the apparatuses of breath—the voice, throat, and diaphragm. At these moments the breathing system's smooth background rhythm for life flares up into an uncontrolled, arrhythmic, spasmodic bursting. Between the unconsciously dominated bi-rhythm of breathing and the semiconsciously controlled off-rhythms and intensities of speaking, we find a whole series of bodily onomatopoeic eruptions—yawwwan, hicccuupp, haaachoo, wahhhhh, haaahaaaha, and aaaaahhhh. Are these the most condensed point of the speaking body, breath, and the preverbal?

Some of these actions develop in utero—like hiccups and yawning. We hiccup most as fetuses, with hiccups declining over the lifespan. Some claim that women hiccup more, though men are more likely to develop intractable hiccups. Charles Osborne has the record for the longest known recorded bout of hiccups, lasting sixty-seven years, between 1922 and 1990. The first time my partner and I went away together, he hiccupped for three days straight. I even sat on the bed of our hotel room and tried to help him with breathing exercises, which he found preposterous. He hid his embarrassment. I found it touching. We eventually parted ways and he hiccupped all the way to Italy, where it promptly abated. It was an auspicious beginning.

A hiccup is an inspiration of breath that is quickly terminated by a closing of the glottis, leading to the infamous "hic" sound. Recall that the hiccup is an aberration going back to when amphibians moved between water breathing and breathing in the air, making it an irruption of an earlier life. Sleep stops hiccups, though breathing of course continues. Why? We don't know a lot about hiccupping. Nor how to cure it. Among the many cures touted for hiccups, orgasm seems to be the most robust.

The psychologist Robert R. Provine in his book *Curious Behavior* investigates the intersections between the old and new in weird yet everyday phenomena such as hiccupping, yawning, tearing-up, laughing, farting, belching, and vomiting. Yawning is interesting. Like sneezing and orgasm, yawning follows a bio-physiological program. This program is old. So old, in fact, that it can override current injuries: people with single-side face paralysis due to stroke can yawn using their entire face. This is because yawning taps into undamaged, unconsciously controlled connections between the brain and motor system.

Freud saw laughter as a form of interruption—a bodily irruption of the unconscious. Human laughter is one of our newer gestures. Chimpanzee laughter is not vocal—more like heaving panting—and scientists speculate that breathing and walking are more synchronized for them than for us bipedal species, where these were disconnected to allow for speaking and thus vocal laughter. Laughter in psychoanalysis is related to the transgressions of cultural mores that provide relief through mutual disinhibition.

Jokes of course tend towards the sexual and aggressive but include all manner of attack on human institutions. Especially on the institution of meaning. In *Inner Experience* the philosopher Georges Bataille said of laughter: "The principle is: if I laugh, the nature of things is laid bare, I understand it, it betrays itself." Interestingly, though, we often laugh before we've consciously understood what is funny—as if our bodies were a step ahead of us.

Some of our curious behaviors interrupt speech, while others seem to work with speech—punctuating it or accentuating it. Provine writes:

> When the ancient and the new, the unconscious and the conscious, compete for the brain's channel of expression, the more modern, conscious mechanism often dominates, suppressing its older, unconscious rival, whether hiccupping or yawning. However, preliminary evidence suggests that hiccupping is not suppressed by the more recently evolved act of speaking, as is laughter; therefore, it does not punctuate the phrase structure of speech. Hiccups, indifferent to grammatical imperative, are distributed more randomly in the speech stream than is laughter.

One woman's hiccups were diagnosed as psychogenic tics since they did in fact punctuate her speech. What we begin to see is a complex interaction between old and new, all the attempts by more recent evolutions to suppress or repress the body's curious behaviors. Then there are the moments they erupt and disrupt.

Some of these eruptions, like laughter, yawning, and crying (but, fortunately, unlike hiccupping and sneezing), are contagious. Freud was fascinated by mass phenomenon, the way we could suddenly start acting less like individuals and more like a multiheaded organism. Mass hysteria often involves the contagion and repetition en masse of bodily fits. There have been epidemics of laughter, screaming, crying, and even dancing.

Yawning, hiccupping, and sneezing involve a massive inspiration of air. Laughing and crying work with expiration, as does coughing. These acts follow a wave pattern that once started must, or perhaps should, reach conclusion. No one wants to be caught mid-yawn. One might speculate that the connection between hiccupping and its ancient evolutionary roots are the reason for so much hiccupping in utero, but this also isn't clear. Hiccups seem to be part of a general spasming of the developing fetal body; the massive spinal cord discharges of energy that scientists believe to be part of the neural development but that could have other functions we don't understand.

It was due to hiccups of the fetus that the neuroscientist Mark Blumberg was recently able to overturn the widely held idea that our bodies twitch during sleep because of our dreams. Blumberg realized that fetuses twitch long before they could plausibly be said to dream. Through MRI imaging, he demonstrated that these twitches came before the mind responded to the dream.

Always, this privileging of the mind—and of course, psychoanalysts are guiltier of that than anyone.

A massive discharge of energy, a change of breath, and a spasm of neural-muscular activity. Our curious behaviors break into speech, they overwhelm us, repeat themselves ad infinitum, marking changes or transitions, hinting at states of arousal and the evocation of memories. For this reason, the feminist critic Catherine Clément sees them as quotidian and catastrophic: "The simple bodily signs that suspend breathing belong to the sphere of syncope; they are recognizable because in their extreme state they can give rise to real fainting fits. So it is with the cough, that banal everyday suffocation; banal, yes, but it is spasmodic, and as such provokes a suspension of being."

Was Reich right about the importance of the spasm? The respiratory movements and bodily-facial expressions of yawning and sneezing are present, to certain degrees, in orgasm: jaw gaping, eyes squinting, stretched facial muscles, head tilted back, tearing, salivating, opening of Eustachian tubes in the middle ear, along with cardiovascular, neuromuscular, diaphragmatic, and respiratory acts, and of course spasm. But I wouldn't follow Reich in arguing for the purposes of the earth-shattering orgasm, nor atmospheric control. Perhaps, like Clément, we ought to think more at the level of the ordinary and the everyday: the shock to the system, the break in the armor. Even if only momentarily, the body in spasm is relieved of its obedience to the mind. We are stretched between agony and ecstasy.

In his book *Breathing: Chaos and Poetry*, the philosopher Franco "Bifo" Berardi turns to the spasm, which, he argues,

can disrupt breathing, the nervous system, and consciousness in a new way: "the spasm is 'chaosmic,' in Guattari's
terms, inasmuch as it invites the organism to remodulate
its vibration and to create, ex-nihilo, a harmonic order by
way of resingularization." Berardi—with Guattari—sees
the breathing spasm as throwing us out of time, especially
already established communal modes of experiencing time
and the rhythms of life.

Spasms thus invite us to step into something singular
and new, as if we both step out of time and back into it,
experiencing it afresh. This would be the possibility of
envisioning another rhythm to life. Berardi's emphasis
on the new is refreshing, flying in the face of New Age
thought understood as a return to some ancient wisdom.
These manifestations of breathing spasms are an energic
disruption.

This constellation has a fascinating antecedent in Plato's
Symposium. Aristophanes, the Greek comedy writer, fails
to give his speech because of a bout of hiccups. Lacan
sees this interruption—this textual hiccup—as the most
important moment in Plato's dialogue, the moment that
changes everything, preempting the later interruption of
the beautiful (and drunk) Alcibiades.

This occurs at a curious moment. Pausanias gives a
rather poor speech that, as Lacan puts it, equates love
with "sheltered capitalization": stocking up and knowing what is worth the investment. In essence: May the
best man win, the rest can fuck off. Nothing about this
way of assessment of love seems divine, let alone true. It
is while Pausanias has paused his speech (which seems
like a joke on his name) that Aristophanes erupts in

hiccups—"Hence," writes Lacan, "it is extremely difficult not to see that, if Aristophanes has the hiccoughs, it's because throughout Pausanias' discourse he's been splitting his sides laughing—and Plato has been doing the same." A sort of commentary.

Ancient Greek advice for the hiccupping person is similar to current wisdom. Aristophanes is told to hold his breath, drink water, sneeze as violently as possible. While Aristophanes tries to cure his hiccups, Eryximachus agrees to speak; he is a medical doctor whose name literally means "battler of belches." Eryximachus believes love strives for perfection in the same way as a body strives for harmony. This talk of bodily harmony is rather funny given the battle with hiccups going on in the background.

Finally, Aristophanes recovers and gives his speech. "Nowhere, not at any moment of the other speeches in the *Symposium* is love taken as seriously or as tragically as it is by Aristophanes," writes Lacan. Aristophanes says that humans are beings who have been cut in two. Panic-stricken, they try to find their other half and perish even if they find them for not being able to come together again. The image of this original, flat, sphere-being with two heads, four arms, and four legs (its genitals joined somewhere inside, but which are resewn onto their fronts when cut) is tragicomic and hideous. Aristophanes is really pushing it, as he does in his plays *The Clouds, The Wasps, The Birds*—all, oddly, figures of the air.

Lacan argues that Aristophanes's speech attacks the spherical as the illusion of wholeness, an image of Oneness and totality. Illusions of oneness are what always lead us astray in love—like that in Pausanias's accumulative theory, and in that of the doctor who sees the body as a harmonizable totality. Instead, Aristophanes eulogizes love in relation to our being split, divided, not-whole. This takes us back to what was originally proclaimed by Phaedrus, earlier in the text, that love involves an asymmetry between the position of the lover and beloved. But what the Gods loved best was when lovers, locked into their respective roles, could switch places.

The *Symposium* is supposed to close with Socrates giving his speech. However, at this moment a drunken Alcibiades bursts in and declares his love for the ugly and old philosopher. Socrates engages in a long play with the beautiful young man, indicating in the end that he does not believe that he really loves him. Socrates refuses to yield to Alcibiades's violent declarations of love. "Through his attitude of refusal, severity, and austerity," writes Lacan, "through his *noli me tangere*, Socrates directs Alcibiades along the pathway of his own good." That is, Socrates holds his place in the name of continued dialogue, like a proto-figure of the psychoanalyst.

For Lacan, the analyst—who no doubt must know a thing or two about love, especially transference love— is like someone who interrupts you and falls down on the ground with hiccups, or who refuses your sexual seductions. The treasure that Alcibiades locates inside of Socrates must be sought elsewhere. While the lover might believe it is in the beloved, and the beloved might believe they contain this object for their lover, Lacan's contention is that it is best sought along the pathways of speech, this modulated breathing. It is the nothing that shines from within in our words—an entirely different way of conceiving of breath.

The long interlude with Alcibiades that ends the *Symposium* feels like a continuation of the refusal of sphericity. Such refusal is tantamount to a transformation of breath. We end with a figure absolutely overtaken by himself, spasming uncontrollably on the floor. And we have another figure who refuses to be undone and unnerved by the nerviest of them all, and who, refusing his advances, invites Alcibiades to speak his truth as best he can. As Michel Serres points out, the *Symposium* is a text full of "burping, hiccoughing, gargling, sneezing, breathing, discoursing (on love: kissing), eating, and drinking." The text is full of noise, wind, belching, and wild movement around what a mouth and breath can do and undo and redo: the unruly scene of human love.

I HAVE READ *Goodnight Moon* to my daughter over and over since she was an infant. Its long, drawn-out goodnight to everything in that surreal green room for a little rabbit in blue striped pajamas. Margaret Wise Brown is a special children's book writer, psychoanalytically inspired, educated at the revolutionary Bank Street School in New York City, where apparently she went too far for even their sensibilities. A *New Yorker* profile notes her tendency toward extremes going all the way back. "She was a tomboy with a terrible temper . . . [W]hen Brown became angry she sometimes held her breath until she turned blue, prompting a nanny to plunge her head into a tub of ice-cold water." Brown's fantastical, wild, and brief life befits the modernist poetics of her writing, hidden in the simplest of stories. She changed children's literature, and, like a good psychoanalyst, she claims she was mere "eye and ear" for the children who were the real writers of her stories.

"Goodnight room, goodnight moon. Goodnight cow jumping over the moon. Goodnight light and the red balloon. Goodnight cow jumping over the moon." At face value, a simple set of rhymes and repetitions: *room, moon, moon, balloon,* and *night, night, night, light, night.* However, Brown manages to evoke the transition from object to sound and image, because who would fail to hear the moo of the cow jumping over the moon, or the transition of night as goodnight light? And who would not think of the balloon in the room, rising in the air, like the moon in the sky outside? The moon outside is shown in the window in the room, while the cow jumping over the moon is a picture in the pictured room.

This is a subtle didactic lesson, to be sure. It speaks the way a child navigates reality as space, air, breath, object, sound, words, jumping from images that are real to imaginary ones in picture books. Also, how these qualities permeate one another, forming a world of associations. All that is seen and named and heard, we must say goodnight to—a version of goodbye—when going to sleep. How does the child know what will be there when it wakes up? The book reassures continuation. Reading the book, night after night, is an enactment of that continuation. Continue reading, continue saying goodnight, continue finding the world still there after your brief absence from it.

The book first lists what there is in the bunny's room, and then takes a second moment to say goodnight to all of it. The clocks only appear in the second round, along with its rhyming partner "socks," which failed to be mentioned in the first half, though they were visually included in the picture. The missing object comes forward to be said goodnight to along with the mention of time. Everything's time will come. In time, everything can be counted.

Brown, whose books only came to be appreciated after her early death at the age of forty-two, was following a new tradition of children's writing that attended to "the here and now" of life rather than fantastical fairy tales. Children find the everyday world magical. They don't need more magic than that. Brown brought her love of language and avant-garde sensibilities to the task of writing children's stories. She says she loves the Oedipal child right before repression sets in. The *New Yorker* profile quotes words from her notebook, "At five we reach a point not to be achieved again," while noting that she elsewhere claimed that children of this age enjoy a "keenness and awareness" that will likely become dampened by life.

Above all I love her love of children's language to the extent of dare—try if you can. She seems to have lived her life this way. The most fantastical addition to the goodnights takes place towards the end of the book. "Goodnight comb, goodnight brush. Goodnight nobody, goodnight mush." It's still a revelation to read this page where *nobody* and *mush* appear and stand firm—worth a goodnight. While mush is certainly food, it is also the dematerialized object, the food eaten, or the food left uneaten. The remnant, there next to nobody and nothing. We are close to the ending: "Goodnight to the old lady whispering hush. Goodnight stars, goodnight air. Goodnight noises everywhere."

We have the final appearance of the only other person next to the little rabbit going to sleep—the old lady and her strange whispers of hush, maybe gentle, maybe harsh. It's hard to tell. And then we zoom out beyond her to the stars, the air, the noise. Every night we say goodnight to all the others. Goodnight to the noises that fill our days with one another, that keep us bound together as humans. Goodnight to the air that carries these sounds, and the stars that mark the outer limits of this world, just like the moon and edges of my room.

It is as if she understands the magical and slightly uncanny journey that a child takes nightly in allowing themselves to go to sleep and wake up again, this rhythmicity like breathing that begins to mark time and time away from other humans. Thinking about this book, about Margaret Wise Brown, I started to see how many children's books reference air, speech, noise, sound. Of course, these books—themselves always read aloud—are the material medium for beginning to learn speech. An induction into another type of breathing, teeming with significance. A space for closeness with words, pictures, sounds, repetitions, at points of transition in the difficult life of a child.

When my daughter is finally left in her crib alone, either at nap time or at night, she speaks and sings. "Crib speech," as it is called: she repeats fragments of songs and books, sometimes interspersed with bits of conversation from the day, close to what Freud, speaking of dream matter, called the day residue, like a fantastical mash-up mix tape. All the phonemic distortions and creative analogizing and agglutinating of words that attack the stability of language is on full display. It really is an early form of free association.

I listen to her follow her mind from the pop of the caterpillar from its egg in the beginning of *The Hungry Caterpillar* to the pop of bubbles in her favorite fish song to *pop pop papa*. She rewrites song lyrics when she reaches a part she doesn't know, and these distortions find their way back when we sing it the next day. "Now I know my ABCs, Mama mama mama me." Language is the place we find ourselves together and then separate.

I try to preserve these inventions rather than remind her of the correct lyrics. We can sing those again later. Years earlier I tried too hard to preserve some of my son's phonetic neologisms, like his substitution of *benge* for *beige*; he said it with such gusto, like Stone Benge. I asked everyone around him to leave it be. He later figured out it wasn't the right word and was furious with me. A mother shouldn't willfully tamper with the symbolic . . . I suppose.

A slightly hysterical distress about these lawful codes could turn into full-blown disgust at the books that were explicitly teaching language to my daughter. They acted like an index of correct transfer from speech to reality. "Besides, they fail to even differentiate what they think they are indicating!" I yelled. Like the word *red* and a picture of a red apple. How am I supposed to know you don't mean *apple*? Don't we learn about language so much better with Margaret Wise Brown, with her gentle free play of associations, than with these didactic monstrosities?

I know better now the chorus of language that can be kept alive both within the constraints of language and by making room for what is so painfully and wonderfully open in it. No poetry is written without a deep sense for the rules, grammar, and structure of language. Willy-nilly invention is cheap. Margaret Wise Brown took a year to write one of her children's books. Apparently, the year was spent refining what was, at first, a furiously scribbled draft on an envelope or napkin.

In *All the World*, which reads like a secular prayer, the author Liz Garton Scanlon also does the work of explaining the turning of day into night, giving its small reader—or listener—a glimpse of a vast world that vanishes in sleep. Of course the "all" of the book is amusing, given the tiny fragment of the world that children know. But it is the word, idea, rhyme, that carries the prayer: the "all," then, is in fact the "all" of language, which we can't see in its totality but that we take in all at once and without consent—which is why Lacan said that language was akin to a trauma, and why, throughout the lives that follow this enormous intake of breath, we forever dream of this "all."

In *All the World*, as the day becomes noon, a storm comes in. "Nest, bird, feather, fly. All the world has got its sky." We see the air and clouds rushing in, birds taking flight on a torrent of wind. It starts the rain: "Slip, trip, stumble, fall. Tip the bucket, spill it all. Better luck another day. All the world goes round this way." This is the page that most will remember, where the skies, filled with rain, gather into a puddle. Time and space collide in the image of weather and the day that goes awry, that stumbles in the same way as children with their little bodies and new instrument of speech.

A little fear and hesitation visit these images, which seem to push us inside, away from the weather. Dinner is being prepared. Day turns to night. "Spreading shadows, setting sun. Crickets, curtains, day is done. A fire takes away the chill. All the world can hold quite still." The clearing sky as dusk falls is the image of stillness from which a light turns on inside a home. The idea of waiting, holding, stillness, foreshadows sleep. Sleep is coming after the turns and tumbles of time and weather and bodies throughout a day.

But not yet! The home erupts with people and sounds. "Nanas, papas, cousins, kin. Piano, harp and violin. Babies passed from neck to knee. All the world is you and me. Everything you hear, smell, see." It's an almost manic moment following the sullen stillness that was atmospheric. The feeling of the baby being passed, the feeling of being close to the neck, or bounced on the knee, the difference in these positions of what you can smell or see.

The rhythm of the book is foretold by the "all" that must be evoked and navigated, from the world and weather to the feelings and smells and sounds at home. This "all" emerges as a kind of joyous human uproar before the ending of the book, which sounds its only false note: "Hope and peace and love and trust. All the world is all of us." If we must, we must.

I thought of these books, the way they interpolate their readers and listeners into the atmosphere of language, while reading Louise Glück's *Marigold and Rose*, the last book she published during her lifetime—her last word. This short fictional work is about her baby twin granddaughters, reading more like a children's story than adult fiction. The twins think about stories and memories and memories of stories they've heard. "She had loved long, long ago (being a twin, she liked things that happened twice) but she had become aware of another way to begin, a way Mother and Father both used when they read stories at bedtime. Once upon a time: that is what the stories said." But there was a sticking point—*upon*. They didn't know what it meant, and it only seemed to be used when reading stories.

"Up, he said, and on. He picked up each twin in turn and held her up (saying the word) and then put her on (usually Mother and Father's bed)," writes Glück. But the twins still can't get a grip on this word. *Once* felt better because Marigold heard the *one* in it, which always began the counting lessons. "And time was the difference between waking up and going to sleep ... Once must mean that time doesn't happen again." Like when you fall, you only fall and hurt yourself once in that one way.

Marigold decides she wants to write a book and settles for *once time* as the way to start, leaving out the confusing *upon*. "She was trying to hear what the book wanted. Then she listened and waited. But the book was completely silent in that way of nonexistent things ... When the book is ready to talk it will talk. Like us, Marigold thought." *Once time* is like *long, long ago*. And the twins were beginning to remember things even though there wasn't much behind them.

In not long at all, they had gone from not breathing and living in the water like tadpoles to being able to breathe and even hold a cup of water and drink by themselves. That's a lot! Marigold wants to learn to remember before she *must* remember—this is what gives her the idea of writing a book. "But actually it was Rose who remembered farther back, being the older twin. I will have to breathe first, Rose thought (it was her first memory). I will have to teach her."

At the end of Glück's book, they must go to sleep after a party. This is no way to end a book, thinks Marigold. This ending is too soon! Marigold, watching Rose sleep, tries to conjure her *long, long ago*, trying to make it real. And then *once time*, as if this could change the ending. But everyone was sleeping. Later, "deep, deep in the night," Rose awakens to see Marigold sleeping. She thinks to herself how she must be having a wonderful dream about her book: "In the dream, Marigold was writing her book, a real book that people who could read would read . . . The end was the morning. I think I must have read that somewhere, Marigold thought, later the next day. But of course she couldn't have since she couldn't read."

I love that Glück turns some of her last moments of writerly attention to a children's story, that her last words return us to first words. Or better, the meta-story of children listening to stories—ones that destine us to become speakers, and maybe writers. Reading aloud is a magical exposure to words that speak to the enduring mystery of life, breath, language, sleep, and time.

In *Goodnight Moon* and in *All the World*, we end with air and the sounds that travel through them as the thread connecting us to our loved ones. The child, I think, must be consoled at sleep. Helped to believe the parent, and everything else in their room and beyond, will still be there in the morning to welcome them to a new day. The incantation of the children's story is the rhythm of bodies and language connected to the rhythm of the day dramatized as the changes in the sky, from light to temperature to weather to seasons. To the stilling of the body after a day's work and play.

Air is so palpably visible in the life of a child: at once written into the most indelible children's stories and the medium through which they're received. How have we forgotten the air? The air is there, all around us, everywhere, and it is gathered into speaking to help the child feel the extension and continuation of all the world. It is a first intimation of time where night is the moment when the sounds stop. Air is part of the sensation of a body in space, and speaking gives meaning as the space children take on in the mind of adults who repeatedly babble with them, read, sing. This means that at one point we were so close to the air. From this vantage point, doesn't it seem like we would have done anything to protect it? Where did this feeling go?

AFTER SPENDING TWO decades with a handful of patients, you have a sense of what someone must do with language. It's not only what they must say to you; it is also what they remember, dream, interrogate, who they encounter in their life, with what words. I still marvel that speaking—simply speaking—can slam you headfirst into your limitations, and, conversely, that boundaries can be shattered with the right word. The time it takes to do this work is certainly a luxury. But who shouldn't be offered what my dear friend, the psychoanalyst Patricia Gherovici, has called the rescue breathing of a listener?

In my years as an analyst, I've come to better understand Freud's skepticism about the human wish for happiness and our ideals of progress in civilization. If this is how long it takes a single person to face reality and to win back a measure of pleasure, what can we hope for the many? Psychoanalysis does have social aspirations. A perspective on the ubiquity of human conflict and illusion is always necessary. The impossibility of loving our neighbors as ourselves—Freud practically speaks about nothing else, says Lacan, adding that besides, most of us hate ourselves.

A neurotic will-to-destructiveness towards our neighbors needs a more protective social net, where Eros can gather more strength. To retranslate Apuleius's telling of the story of Eros and Psyche, "for there is no place for desire to dwell, save the breath, who animates all things; and there is no meaning for the breath to live and be awake, save for the sake of desire."

Community appears differently to the psychoanalyst in me. I don't want communion or oneness or mass liberation. If I found these appealing at one time (dreams of starting a commune with friends throughout my twenties), the difficulty of these sentiments when listening to patients has steered me otherwise. Maybe sideways, to follow Berlant. Air, which connects all people while measuring the distance between them. These investigations of breathing have been a long battle with the oceanic wish: I needed a seawall that would still allow me to swim. Breathing, writes Rilke, "the counterpoint to my own rhythm, a single wave by which I gradually become the sea."

I want to develop our human rhythms, vary them, push and play with them, before dissolving them into a single wave. What does it take to discover a feeling for time that isn't the incessant rush of the clock? It isn't easy to create space in life, you who are so guilty of being a selfish navel gazer. "Immensity is within ourselves," writes Bachelard in *The Poetics of Space*. "It is attached to a sort of expansion of being that life curbs and caution arrests, but which starts again when we are alone."

Breathing is the "inspiration and expiration of being," writes Maurice Merleau-Ponty. We must begin to know something intimate about how a soul cracks when it meets with the world. Did you know that when an infant takes its first breath, an opening will close between the two compartments of its heart that, in the months leading up to that moment, had oxygenated its blood with its mother's?

Psychoanalytic victories tend to be marginal achievements, not to mention achievements at the margin of the social. An intransigent aloneness after being in psychoanalysis is inescapable. Maybe *solitude* is a better word. The analysts I know whisper about it together. Groups? Ha! This isn't nihilism. There is a wonder for our inventiveness as a species, even for our capacity to invent so many species of unhappiness. A look upon the world askance allows for an ironic compassion much needed in such catastrophic times.

Patients do not talk to me because I'm such a virtuoso at the art of conversation (I'm certainly not), nor because I provide such unique answers or solutions (I certainly don't). What I have learned is to hold the place of listener, and from that place to remain on the lookout for an opening in their speech that is elided by day-to-day life or the confines of communication. Suddenly they think through their questions differently than, for all their ruminating, they'd been able to before—as if some miraculous breathing room has been found in a deep pocket of their lungs.

Allow me to speak one last time about my breathing, this time not because of asthma but because of my difficulties with speaking. A whole scene unseen, a hidden breakdown, could be ascribed to a confrontation with speech—importantly the other's speaking. I hid this from myself even as I made it my chosen profession. A compromise to be sure, but to be effective as a psychoanalyst I would have to come to grips with a dimension of speech that was maddening to me, that rendered me utterly helpless, in a state of free fall. As the writer Vernon Lee says in *Music and Its Lovers*, "[music] lives in our breath, yet it seems to come from a distant land which we shall never see, and to tell us things we shall never know."

For a long time, I had a consistent way of ignoring what was deadly or aggressive in a person's way of speaking. I might react hysterically—fly off the handle, black out, grow aggressive in a way that left me the guilty party. But I could never remember what had happened. Something was un-hearable. No words for an upset beyond distaste for certain behaviors. The self-righteous charge against the narcissists that captured nothing.

When I couldn't ignore this speaking any longer, I thought I could demand that it stop by will alone. Or at the very least a clever argument could make it go away. You can neither ignore, explain, nor stop it. You can only work with it or wait. I could be reduced to a child with their hands over their ears. Yelling la-la-la-la-la-la. I put songs on repeat—sometimes played for weeks—to erase whatever it was. I still couldn't let myself know. I had to erase an erasure with more sound? How had I become a professional listener?

Like any beautiful soul, what is bemoaned is close to heart. In tense moments there was something I could do with my voice that stopped the other's speech that

I didn't want. I could arrest them in their tracks with a kind of booming ferocity. But this maneuver prevented any further talking. Only rarely did we make it back to the point of disagreement. Where did I learn this? I had never done it as a child. I never screamed or fought back. I must have fashioned this armor slowly against an enemy in my dreams. Motherless Athena born from the forehead of Zeus.

A rather commonplace anxiety around public speaking emerged in college. I could sometimes be one of those people in a classroom that was completely silent for the entire year. Other times I was fine, even vociferous. Looking back, a powerful transference to a professor and a feeling of mutual respect was needed. In my teachers, those adults who I wanted to listen to, I found a life raft—yet it only exacerbated the sinking feeling of having to endure speech I didn't want to hear.

This mysterious inhibition ran counter to an image I had cultivated of being outspoken. Saying I had trouble speaking came as a surprise to others, sometimes even to myself. I was able to hide this anxiety when I did speak with lackadaisical posturing, as if I didn't have a care in the world. But my heart would be racing, and I would go slightly deaf as a warm sensation rose up the nape of my neck. It would take a minute before I came to after speaking through this anxious haze.

To speak through anxiety, you must allow yourself to begin delivering what you want to say without hearing what you are saying, without really being there. Would I ever be able to speak extemporaneously? Not memorization and delivery, but preparation and true spontaneity, casting my notes aside? Most of Jacques Lacan's work consists of transcripts from his seminars where he would speak for hours, surprising even himself, a kind of *how about that* as a mysterious logic unfolded before your eyes.

Old-fashioned psychoanalysis could surmise underneath my inhibitions something pent-up, something repressed, that wants to get said. The person who can't speak is afraid of screaming or cursing or telling some secret. Some violent or sexual confrontation whose image is of people arguing or where speaking intensely is feared. Maybe an exhibitionistic wish is being shoved down by the symptom.

Freud once described a hysterical patient who had a fear of seeing people talking in an excited manner inside shops and developed agoraphobia. In his case known as the Ratman, Freud's patient recalled lambasting his father as a child with the only words he knew—you plate, you table, you lamp—and stopped him mid-spanking. His father uttered the powerful words, "He'll either be a great man or a criminal," words that seemed to mark him symptomatically. His intense libidinal speech had created a bifurcated prophesy that made him doubt all his decisions.

In fact, in most of Freud's cases you'll find a scene of speech, like when his young hysterical patient Dora was told by her father's mistress's husband, "My wife means nothing to me," and she suddenly slapped him. Because she loved this woman, she had tolerated the strange situation of having to tend to these others on behalf of her father. The words, like a release from a magical spell, lifted a veil after which she overturned the entire apple cart.

An ocean of silence in my past appeared before me. I was, it is true, a quiet child. I didn't say much. I didn't complain. I didn't speak really. I wasn't fundamentally there. As Kathy Acker writes in *Empire of the Senseless*, "For a long time I had remained apathetic. So sure that my words meant nothing to anyone that I no longer spoke unless circumstances forced me to." In sixth grade, I didn't make it into the school play—more public speaking—and my teachers (surmising that something was off) asked to talk to me. I put my head down on the desk and wouldn't look at them. They were trying to be kind; they didn't realize they had touched something fundamental.

This question of speaking was funneled into some childhood obsession with parrots. I wanted a baby hyacinth macaw that I could teach to talk. I was keenly aware that this is what a parrot can do that no other pet can. There is a yellow accent at the eyes and mouth of the otherwise monochromatic deep blue of the bird. I see some proto-image of myself. My eyes staring at the other's mouth. My mouth fearing the gaze of the other. The gaze and voice are the most immaterial of the erogenous zones whose objects vaporize into thin air.

I would ride my bike to The Parrot Jungle, a tourist attraction near my house, where I would marvel at the birds and talk to them. They were awaiting my visits, I imagined. I looked longingly at them on their perches from the onsite cafeteria.

A parrot that we had when I was a child used to scream the name of my nanny exactly as I screamed it—Emmmmmmmaaaa! It echoed through the neighborhood. Another peculiarity of my life: I was raised by a nanny who didn't speak English and who was only educated until the age of seven in the Philippines. She had false teeth that she took out every night and put in a glass next to our bed on the floor, which made her even more incomprehensible. The joke (not funny) was that I was the only person who could understand her.

I witnessed a lot of arguing, a lot of vicious speech as a child. I was a child who was talked at. My maternal grandmother used to say my father had diarrhea of the mouth. He talked until you surrendered. This condemnation was part of her verbally assessing most people that she knew. I was caught in the crossfires of speech that condemned other speech, speech that diagnosed other people's speaking. None of it speaking *to*. The makings of a psychoanalyst.

I also developed the habit of going dead in the face of my father's talking. I would put my head down, curl into a ball, and close my eyes until he stopped. I must have wondered why he kept going, even with this display of finding something unbearable, of not listening and disappearing. But I don't remember wondering. I only remember having extreme reactions when he came home from his trips.

I began to surmise that I carried around a question about those who don't stop talking when you've already stopped listening, realizing it was my lightning rod. Physiologically electric—I was, as they say about patients in the hospital, agitated. Also, utterly impatient. The opposite reaction could also manifest with force. In the face of someone whose speech I wanted to hear, I was immediately and dangerously in love.

It's important to understand that despite remembering all of this, I could neither remember having a question about being listened to, nor could I remember my father's voice. I couldn't hear his tone, his way of speaking—yet I knew it drove me crazy. Why was this so repressed?

I searched for speech worth listening to. I loved speech that was subtle, careful, attentive to its listener, rarely demonstrative. When necessary, I liked speech that could strike hard, with surgical concision. Speech could be wild and messy, rambling and digressive, if its listener was kept in mind. There had to be an eye for timing and tact. What an ask! You are what I want to hear?

This search for an ideal is both a capacity and a defense. The capacity is a large part of whatever abilities I have as a writer and teacher. But it is backed by a monstrous scene of judgment, a defensive wall that hadn't been broached. At least not yet, or not then.

One day in my twenties, I found myself having to force my mother to back down during one of her verbal tirades. With all my strength, I calmly said that she had misunderstood me but that she seemed incapable of hearing that from me. Either she could stop, or one of us would have to leave. She left.

My hands were shaking. My entire body seemed to shut down after the confrontation. What I said wasn't much, but it was a line I had never been able to draw. We never had this problem again. It was separation despite a life of having lived very separately. Winnicott would say that we found out that we could survive destruction. I would emphasize the verbal destruction. The possibility of joking with my mother appeared for the first time: our new terrain of loving one another.

I love listening to my children. I sometimes worry that I demanded that they cultivate speech that pacified me since I'm so sensitive to it; I do my best to listen carefully to them, speak *with* them. I remember when my son tried out verbally excoriating me for the first time; I had egged him on to give me his best shot. He succeeded. I saw that he saw I was impressed.

As toddlers my children had similar idiosyncratic phrases of warding off the Other. My son used to yell, "I want to leave alone." My daughter says, "Go away. Leave my lone. It's my lone. Not your lone." Stopping the Other's intrusion is close to the verbal threat of leaving. It's a fascinating condensation of what I was confused about as a child: Was I abandoned, or should I have been left more alone? If we can say this to each other, doesn't it mean we don't have to leave?

While many think of psychoanalysis as the excavation of trauma—sexual and aggressive scenes that overwhelm—the dimension of breathing and speaking has been a subtle, subterranean presence. Breathing takes us to this unpredictable kernel of speech, the breaking point between body and world. For Lacan language is *the* primal scene. Even when it comes to real events there is always what people say, what they don't say, and what is unsayable that we must negotiate.

I think we all forget that, when young, we were terrified by all those adult people speaking. Charlie Brown knee height and those voices going WAHWAHWAH. Having just learned to speak, you are close to how little is *ever* understood of what others say, close to the question of why people say what they say—especially to you. Children also have a raw sense of what is forbidden to say, what must be said to appease. This excess in the exchange of words, we simply can't quite get our heads around.

Lacan elevated the poetic, musical, and tonal qualities present in speaking. The analyst needs to have the widest repertoire around language ready to hand. Relinquishing control over speaking also dethrones the task of understanding. Let the speech pour out of you. Let this torrent of air blow over you.

Aren't symptoms embodied by a relationship to speech the heart of psychoanalysis? Isn't this human sensitivity what we see so powerfully in psychosis that can't figure out how to stop hearing something monstrous about voices? In my time on the couch, I have come to know speech that is blocked. Helpless speech. Speech that wants a freedom it can't yet have. Speech that is shorn of libido. Repetitive speech. Speech that is marvelously abstract. Speech that is excessively labile.

There is erotic and sensual speech. Speech that gets lost in detail and digression. A maze! Speech that degenerates into crying. Speech that is always screaming. Furious speech. Speech caught in understanding. Speech that never feels finished. Speech that happens too fast. Speech filled to the brink with demands. Speech that never reaches itself. Wanting to stop speaking. Not speaking.

I developed my only new symptom in analysis during these language lessons—a tic where I would hear things at a slight lag, respond with an even longer lag, which made people (especially certain loved others) repeat things to me. I didn't want to hear what was said the first time, no less a second or third. I wanted to scream stop. Scream that I always hear! Who was I was screaming at?

In retrospect, I embodied all sides of my nightmare, both taken for not listening and needing to be talked at. From terror to horror and wonder, it's right there on the tip of your tongue. My early childhood bodily symptom of asthma felt as if it dissolved into this Janus-face of language: language filled with anxiety and ecstasy, care and carelessness, a perturbance of life and a deadly abyss. I am lucky psychoanalysis took an original symptom—a real bodily diffusion—and battled with the multiheaded beast, shoving me into this difficult scene of speaking.

Darian Leader in his book *Why Do People Get Ill?* notes that our distress needs the minimal idea of an addressee. "Could this be why asthma sufferers fare better if they are encouraged to write about their troubling experiences?" It would not seem unreasonable, he concludes, that the abolition of human dialogue influences immune functioning. Perhaps the overactive immune functioning of asthmatics, to say nothing of so many chronic conditions, is a call to the other asking for entry to the scene of speaking.

Isn't a younger generation calling for some change around the way we speak? We have demands for safe spaces, trigger warnings, radical inclusivity, fast affirmative talking, will-to-complain and be heard, community as online connectedness through writing. Even the surge of interest in autism—the disorder that limits the social aspects of language—circles this wish. Language feels precarious to my students—a site of maximal vulnerability that feels connected to their sense for the fragility of the environment and certain segments of the population.

My nervous system gets overwhelmed, they tell me. I hear them asking for something less brutal while struggling with their own violent reactions. These questions arrive for them so much sooner than they arrived for me. I can't help but wonder what it means, this striving for an environment of speech I want to hear, this fantasy now made real by today's algorithms. I can't help but hear in their concern the weight of responsibility they inherit for the damage we have done to the earth, to the seas, and to the air.

My daughter will be generation alpha. Starting at the beginning of the alphabet. Rewriting it all again. I think of my analyst, closer in age to my father. The silent generation. She stayed true to form—mostly, she was silent. She never tried too hard to understand, and in my attempts to understand myself—doctor, heal thyself—she sometimes showed concern. Maybe (I intimate) a little pity. I saw her grow old. Did she understand me?

"The wish to be understood," writes Adam Phillips, "may be our most vengeful demand, may be the way we hang on, as adults, to our grudge against our mothers; the way we never let our mothers off the hook for their not meeting our every need. Wanting to be understood, as adults, can be our most violent form of nostalgia." My wish for a more perfect womb.

I meet the rush of cold air.

Clarice Lispector writes in *The Passion According to G.H.*, "In the interstices of primordial matter is the line of mystery and fire that is the breathing of the world.

The world's continual breathing is what we hear and call silence." Language cannot say it all. We are still learning what we can do with language. I revere the fundamental rule. What an ask. Say anything, say everything you can say. And then silence of breath, abyss of air.

ACKNOWLEDGMENTS

Writers have a new symptom—reminiscing. They speak of some mythic creature called a "real editor" that appears to have become extinct. These editors, so they say, know about writing, which is good because authors don't know what they are doing—writing lives in a cloud of unknowing—and the real editor's support is critical. They know how to push, stay quiet, wait, rewrite, send you something to read, ask questions, know what is finished. They see where a book is going before the writer does, guiding them to the finish. At some points, it is as if the writing happens together. Woe, that such a creature has been lost! To my great surprise, I found out that editors aren't extinct, just an endangered species.

Will Rees would never let me write such a ridiculous paragraph. What an absolute pleasure it was working with him on this book. It feels fortuitous that we share a passion for neurotic illness, cruel optimism, and air. My gratitude to Kendall Storey and the team at Catapult for believing in this project, and their careful editorial eye. The freedom she accorded me is rare.

I also must thank my research assistant Kasen Scharmann. One night he asked me if I first wanted to be a psychoanalyst or a writer. I hesitated answering. I have always felt like I wanted to be a psychoanalyst for as long as I could remember, but I also had a strange memory of thinking I was too scared to be a writer. I knew it, he said. You like writing too much. Kasen happens to be an excellent philosopher and writer, so reading for me, and researching every whim from a constant dialogue we had throughout the writing of this book, has left a million traces across these pages.

Also, my deepest gratitude to Stacy Ruttenberg, whose exuberance for questions about breathing and the body, the strange details of any psychoanalytic case or social phenomenon, was a comfort every time I needed to find some inspiration. She is a nurse, training to be a psychoanalyst, and her singular know-how was invaluable.

I want to thank my best friend and colleague, Patricia Gherovici. We have written together on breathing and psychoanalysis, and our conversations and work were the foundation that this book was built on. My gratitude to my colleagues: Eliana Betancourt, Marcus Coelen, Josh Cohen, Edward Dioguardi, Fiona Duncan, Bracha L. Ettinger, Loryn Hatch, Matt Johnson, Darian Leader, David Lichtenstein, Elissa Marder, Kerry Moore, Julie Beth Napolin, Jesse Pearson, Eyal Rozmarin, Vaia Tsolas, Astra Taylor, Rachel Valinsky, Camilla Wills, Eleanor Ivory Weber.

And last, to those with whom I share the most intimate air—this book, for you, with love, to Richard, Alma, Soren, and Isobel.

WORKS CITED

Acker, Kathy. *Empire of the Senseless*. Grove Press, 1988.

Albano, Caterina. *Out of Breath: Vulnerability of Air in Contemporary Art*. University of Minnesota Press, 2020.

Albaret, Céleste. *Monsieur Proust*. New York Review Books Classics, 2003.

Bachelard, Gaston. *Air and Dreams: An Essay on the Imagination of Movement*. Translated by Edith R. Farrell and C. Frederick Farrell. *Dallas Institute Publications*, 2011.

Bachelard, Gaston. *The Poetics of Space*. Translated by Maria Jolas. Beacon Press, 1994.

Bachmann, Ingeborg. *The Book of Franza and Requiem for Fanny Goldmann*. Translated by Peter Filkins. Northwestern University Press, 2010.

Bachmann, Ingeborg. *Malina*. Translated by Philip Boehm. New Directions, 2019.

Bachmann, Ingeborg. *The Thirtieth Year*. Translated by Mary Gilbert. Holmes & Meier, 1995.

Baldwin, James. "The New Lost Generation" (1961). *James Baldwin: Collected Essays*. Library of America, 1998.

Bataille, Georges. *Inner Experience*. Translated by Leslie Anne Boldt. State University of New York Press, 1988.

Beckett, Samuel. *Breath*. In *The Complete Dramatic Works of Samuel Beckett*. Faber & Faber, 2006.

Beckett, Samuel. *Waiting for Godot*. In *The Complete Dramatic Works of Samuel Beckett*. Faber & Faber, 2006.

Berardi, Franco "Bifo." *Breathing: Chaos and Poetry*. Semiotext(e), 2019.

Berlant, Lauren. *Cruel Optimism*. Duke University Press, 2011.

Berlant, Lauren. "Slow Death (Sovereignty, Obesity, Lateral Agency)." *Critical Inquiry* 33, no. 4 (2007): 754–80.

Bersani, Leo. *Receptive Bodies*. University of Chicago Press, 2018.

Bion, Wilfred R. *Bion in New York and Sao Paulo: And Three Tavistock Seminars*. Karnak Books, 2019.

Bion, Wilfred R. *Four Discussions with W. R. Bion*. Karnac Books, 2019.

Bion, Wilfred R. *Learning from Experience*. Routledge, 2023.

Burroughs, William S. *The Electronic Revolution*. Expanded Media Editions, 1970.

Butler, Judith. *What World Is This?: A Pandemic Phenomenology*. Columbia University Press, 2022.

Canetti, Elias. *The Conscience of Words*. Translated by Joachim Neugroschel. Farrar, Straus and Giroux, 1984.

Carel, Havi. "Invisible Suffering: The Experience of Breathlessness." In *Atmospheres of Breathing*, edited by Lenart Škof and Petri Berndtson. State University of New York Press, 2018.

Carson, Anne. "The Gender of Sound." In *Glass, Irony, and God*. New Directions, 1995.

Carson, Anne. *Men in the Off Hours*. Vintage, 2001

Celan, Paul. *The Meridian: Final Version—Drafts—Materials*. Edited by Bernhard Böschenstein and Heino Schmull. Translated by Pierre Joris. Stanford University Press, 2011.

Cioran, E. M. *All Gall Is Divided: Gnomes and Apothegms*. Translated by Richard Howard. Arcade Publishing, 1999.

Clark, Robert. "Cosmic Consciousness in Catatonic Schizophrenia." *Psychoanalytic Review* 33, no. 4 (1946): 460–504.

Clément, Catherine. *Syncope: The Philosophy of Rapture*. Translated by Sally O'Driscoll and Deirdre M. Mahoney. University of Minnesota Press, 1994.

Coccia, Emanuele. *The Life of Plants: A Metaphysics of Mixture*. Polity Press, 2019.

Connor, Steven. *The Matter of Air: Science and Art of the Ethereal*. Reaktion Books, 2013.

Corruccini, Robert, Louisa B. Flander, and Samvit S. Kaul. "Mouth Breathing, Occlusion, and Modernization in a North Indian Population: An Epidemiological Study." *Angle Orthodontist* 55, no. 3 (1985): 190–96.

Crawley, Ashon T. *Blackpentecostal Breath: The Aesthetics of Possibility*. Fordham University Press, 2017.

Daston, Lorraine. *Against Nature*. MIT Press, 2019.

de Saussure, Ferdinand. *Course in General Linguistics*. Edited by Charles Bally, Albert Sechehaye, and Albert Riedlinger. Translated by Wade Baskin. McGraw-Hill, 1966.

Diehm, Cade. "Too Late for Earth, Too Soon for the Stars." *Ursula*, September 22, 2023.

Ettinger, Bracha L. *Matrixial Subjectivity, Aesthetics, Ethics*. Vol. 1, *1990–2000*, edited by Griselda Pollock. Palgrave Macmillan, 2020.

Fanon, Frantz. *The Wretched of the Earth*. Translated by Richard Philcox. Grove Press, 2004.

Fenichel, Otto. "Respiratory Introjection." In *The Collected Papers of Otto Fenichel*, edited by Hannah Fenichel and David Rapaport. W. W. Norton, 1954.

Ferenczi, Sándor. *The Clinical Diary of Sándor Ferenczi*. Edited by Judith Dupont. Translated by Michael Balint and Nicola Zarday Jackson. Harvard University Press, 1995.

Ferenczi, Sándor. *Thalassa: A Theory of Genitality*. Routledge, 1989.

Ferenczi, Sándor. "The Unwelcome Child and His Death-Instinct." *International Journal of Psycho-Analysis* 10 (1929): 125–29.

Freud, Sigmund. *Beyond the Pleasure Principle* (1920). In *1920–1922*, vol. 18 of *The Standard Edition of the Complete Psychological Works of Sigmund Freud*, edited and translated by James Strachey. Hogarth Press and the Institute of Psycho-Analysis, 1953.

Freud, Sigmund. *Civilization and Its Discontents* (1930). In *1927–1931*, vol. 21 of *The Standard Edition of the Complete Psychological Works of Sigmund Freud*, edited and translated by James Strachey. Hogarth Press and the Institute of Psycho-Analysis, 1953.

Freud, Sigmund. "A Disturbance of Memory on the Acropolis" (1936). In *1932–1936*, vol. 22 of *The Standard Edition of the Complete Psychological Works of Sigmund Freud*, edited and translated by James Strachey. Hogarth Press and the Institute of Psycho-Analysis, 1953.

Freud, Sigmund. "[Drives] and Their Vicissitudes" (1915). In *1914–1916*, vol. 14 of *The Standard Edition of the Complete Psychological Works of Sigmund Freud*, edited and translated by James Strachey. Hogarth Press and the Institute of Psycho-Analysis, 1953.

Freud, Sigmund. *From the History of an Infantile Neurosis* (1918). In *1917–1919*, vol. 17 of *The Standard Edition of the Complete Psychological Works of Sigmund Freud*, edited and translated by James Strachey. Hogarth Press and the Institute of Psycho-Analysis, 1953.

Freud, Sigmund. *The Future of an Illusion* (1927). In *1927–1931*, vol. 21 of *The Standard Edition of the Complete Psychological Works of Sigmund Freud*, edited and translated by James Strachey. Hogarth Press and the Institute of Psycho-Analysis, 1953.

Freud, Sigmund. *Inhibitions, Symptoms and Anxiety* (1926). In *1925–1926*, vol. 20 of *The Standard Edition of the Complete Psychological Works of Sigmund Freud*, edited and translated by James Strachey. Hogarth Press and the Institute of Psycho-Analysis, 1953.

Freud, Sigmund. *Introductory Lectures on Psychoanalysis* (1917). In *1915–1916*, vol. 15 of *The Standard Edition of the Complete Psychological Works of Sigmund Freud*, edited and translated by James Strachey. Hogarth Press and the Institute of Psycho-Analysis, 1953.

Freud, Sigmund. *Letters of Sigmund Freud*. Edited by Ernst L. Freud. Translated by Tania Stern and James Stern. Basic Books, 1960.

Freud, Sigmund. *Moses and Monotheism* (1939). In *1937–1939*, vol. 23 of *The Standard Edition of the Complete Psychological Works of Sigmund Freud*, edited and translated by James Strachey. Hogarth Press and the Institute of Psycho-Analysis, 1953.

Freud, Sigmund. "Obsessions and Phobias" in *1893–1899*, vol. 3 of *The Standard Edition of the Complete Psychological Works of Sigmund Freud*, edited and translated by James Strachey. Hogarth Press and the Institute of Psycho-Analysis, 1953.

Freud, Sigmund. *A Phylogenetic Fantasy: Overview of the Transference Neuroses*. Edited by Ilse Grubrich-Simitis. Translated by Axel Hoffer and Peter T. Hoffer. Belknap Press, 1987.

Freud, Sigmund. "Some Character-Types Met with in Psycho-Analytic Work" (1916). In *1914–1916*, vol. 14 of *The Standard Edition of the Complete Psychological Works of Sigmund Freud*, edited and translated by James Strachey. Hogarth Press and the Institute of Psycho-Analysis, 1953.

Freud, Sigmund. *Studies in Hysteria* (1895). In *1893–1895*, vol. 2 of *The Standard Edition of the Complete Psychological Works of Sigmund Freud*, edited and translated by James Strachey. Hogarth Press and the Institute of Psycho-Analysis, 1953.

Freud, Sigmund. *Three Essays on the Theory of Sexuality* (1905). In *1901–1905*, vol. 7 of *The Standard Edition of the Complete Psychological Works of Sigmund Freud*, edited and translated by James Strachey. Hogarth Press and the Institute of Psycho-Analysis, 1953.

Ghorayshi, Azeen. "After Antidepressants, a Loss of Sexuality." *New York Times*, November 9, 2023.

Glück, Louise. "Marigold and Rose." *Poems: 1962–2012*. Farrar, Straus and Giroux, 2012.

Gornick, Vivian. *Fierce Attachments: A Memoir*. Beacon Press, 1987.

Guattari, Félix. *Chaosmosis: An Ethico-Aesthetic Paradigm*. Translated by Paul Bains and Julian Pefanis. Indiana University Press, 1995.

Gumbs, Alexis Pauline. "Undrowned: Black Feminist Lessons from Marine Mammals." *Soundings* 78 (2021): 20–37.

Holmes, Anna. "The Radical Woman Behind 'Goodnight Moon.'" *New Yorker*, January 31, 2022.

Irigaray, Luce. *The Forgetting of Air in Martin Heidegger*. Translated by Mary Beth Mader. University of Texas Press, 1999.

Irigaray, Luce. "From the Forgetting of Air to To Be Two." In *Feminist Interpretations of Martin Heidegger*, edited by Nancy J. Holland and Patricia Huntington. Penn State University Press, 2001.

James, William. *William James: Writings 1902–1910*. Library of America, 1988.

Kafka, Franz. *The Complete Stories*. Schocken, 1987.

Kierkegaard, Søren. *The Sickness Unto Death: A Christian Psychological Exposition for Upbuilding and Awakening*. Edited and translated by Howard V. Hong and Edna H. Hong. Princeton University Press, 1983.

Klein, Melanie. *Love, Guilt and Reparation and Other Works 1921–1945*. Free Press, 1975.

Klein, Melanie. *Narrative of a Child Analysis: The Conduct of the Psycho-Analysis of Children As Seen in the Treatment of a Ten-Year-Old Boy*. Delacorte Press, 1975.

Klein, Melanie. "The Oedipus Complex in the Light of Early Anxieties." *International Journal of Psychoanalysis* 26 (1945): 11–33.

Klein, Naomi. *This Changes Everything: Capitalism vs. The Climate*. Simon & Schuster, 2014.

Lacan, Jacques. *Anxiety: The Seminar of Jacques Lacan, Book X*. Edited by Jacques-Alain Miller. Translated by A. R. Price. Polity Press, 2014.

Lacan, Jacques. *Formations of the Unconscious: The Seminar of Jacques Lacan, Book V*. Edited by Jacques-Alain Miller. Translated by Russell Grigg. Polity Press, 2017.

Lacan, Jacques. "The Instance of the Letter in the Unconscious, or Reason Since Freud" (1966). In *Écrits: The First Complete Edition in English*, translated by Bruce Fink. W. W. Norton, 2006.

Lacan, Jacques. "La troisième." In *Autres écrits*, edited by Jacques-Alain Miller. Seuil, 2001.

Lacan, Jacques. "Lituraterre." In *Autres écrits*, edited by Jacques-Alain Miller. Seuil, 2001.

Lacan, Jacques. "Logical Time and the Assertion of Anticipated Certainty" (1945). In *Écrits: The First Complete Edition in English*, translated by Bruce Fink. W. W. Norton, 2006.

Lacan, Jacques. *The Other Side of Psychoanalysis: The Seminar of Jacques Lacan, Book XVII*. Edited by Jacques-Alain Miller. Translated by Russell Grigg. W. W. Norton, 2007.

Lacan, Jacques. *The Psychoses: The Seminar of Jacques Lacan, Book III*. Edited by Jacques-Alain Miller. Translated by Russell Grigg. W. W. Norton, 1993.

Lacan, Jacques. "The Subversion of the Subject and the Dialectic of Desire in the Freudian Unconscious" (1960). In *Écrits: The First Complete Edition in English*, translated by Bruce Fink. W. W. Norton, 2006.

Lacan, Jacques. *Transference: The Seminar of Jacques Lacan, Book VIII*. Edited by Jacques-Alain Miller. Translated by Bruce Fink. Polity Press. 2015.

Leader, Darian, and David Corfield. *Why Do People Get Ill?: Exploring the Mind-Body Connection*. Hamish Hamilton, 2007.

Lee, Vernon. *Music and Its Lovers: An Empirical Study of Emotional and Imaginative Responses to Music.* Thomas Press, 2007.

Lerner, Ben. "Café Loup." *New Yorker*, August 29, 2022.

Levinas, Emmanuel. *Otherwise than Being, or Beyond Essence.* Translated by Alphonso Lingis. Duquesne University Press, 1998.

Lispector, Clarice. *The Passion According to G.H.* Translated by Idra Novey. New Directions, 2012.

Lockwood, Patricia. *No One Is Talking About This: A Novel.* Riverhead Books, 2021.

Loraux, Nicole. *Mothers in Mourning.* Translated by Corinne Pache. Cornell University Press, 1998.

Loraux, Nicole. *Tragic Ways of Killing a Woman.* Translated by Anthony Forster. Harvard University Press, 1991.

Mahler, Margaret. "Separation-Individuation Process in Children." Produced by the Margaret S. Mahler Child Development Foundation, The Psychoanalytic Study of the Child, 1975.

Marcuse, Herbert. *One-Dimensional Man: Studies in the Ideology of Advanced Industrial Society.* Beacon Press, 1991.

Marder, Elissa. "The Shadow of the Eco: Denial and Climate Change." *Philosophy & Social Criticism* 49, no. 2 (2023): 139–150.

Marx, Karl, and Frederick Engels. *The Communist Manifesto.* International Publishers, 2014.

Mbembe, Achille. "The Universal Right to Breathe." Translated by Carolyn Shread. *Critical Inquiry* 47, no. 2 (2021): S58–S62.

Melville, Herman. *Moby Dick; or The White Whale.* Dodd, Mead, 1922.

Merleau-Ponty, Maurice. *Phenomenology of Perception.* Translated by Donald A. Landes. Routledge, 2013.

Merleau-Ponty, Maurice. *The Primacy of Perception: And Other Essays on Phenomenological Psychology, the Philosophy of Art, History and Politics.* Translated by William Cobb. Edited by James M. Edie. Northwestern University Press, 1964.

Nancy, Jean-Luc. "Communovirus." In *Coronavirus, Psychoanalysis, and Philosophy: Conversations on Pandemics, Politics and Society*, edited by Fernando Castrillón and Thomas Marchevsky. Routledge, 2021.

Nersessian, Anahid. *Keats's Odes: A Lover's Discourse*. University of Chicago Press, 2021.

Nestor, James. *Breath: The New Science of a Lost Art*. Riverhead Books, 2020.

Nieuwenhuis, Marijn. "Porous Skin: Breathing Through the Prism of the Holey Body." *Emotion, Space and Society* 33 (2019).

Orenstein, Peggy. "The Troubling Trend in Teenage Sex." *New York Times*, April 12, 2024.

Phillips, Adam. *Missing Out: In Praise of the Unlived Life*. Farrar, Straus and Giroux, 2012.

Plato. *Symposium*, translated by Alexander Nehamas and Paul Woodruff. In *Plato: Complete Works*, edited by John M. Cooper. Hackett Publishing, 1997.

Provine, Robert R. *Curious Behavior: Yawning, Laughing, Hiccupping, and Beyond*. Belknap Press, 2014.

Puri, Sunita. "The Hidden Harms of CPR." *New Yorker*, August 5, 2023.

Rank, Otto. *The Trauma of Birth*. Dover Publications, 1994.

Rank, Otto. "The Trauma of Birth in Its Importance for Psychoanalytic Therapy." *Psychoanalytic Review* 11, no. 3 (1924): 241–45.

Reich, Wilhelm. *Character Analysis*. Translated by Vincent R. Carfagno. Farrar, Straus and Giroux, 1980.

Reich, Wilhelm. *Function of the Orgasm: Vol 1 of The Discovery of Orgone*. Translated by Vincent R. Carfagno. Farrar, Straus and Giroux, 1986.

Reich, Wilhelm. *Listen, Little Man!* Translated by Ralph Manheim. Farrar, Straus and Giroux, 1974.

Reich, Wilhelm. *Reich Speaks of Freud: Wilhelm Reich Discusses His Work and His Relationship with Sigmund Freud*. Edited by Mary Higgins and Chester M. Raphael. First Noonday Press, 1968.

Rhode, Maria. "Autistic Breathing." *Journal of Child Psychotherapy* 20, no. 1 (1994): 25–41.

Rilke, Rainer Maria. *Duino Elegies & The Sonnets to Orpheus: A Dual-Language Edition*. Edited and translated by Stephen Mitchell. Vintage Books, 2009.

Ross, Alex. "What Is Noise?" *New Yorker*, April 15, 2024.

Ruddick, Bruce. "Colds and Respiratory Introjection." *International Journal of Psychoanalysis* 44 (1963): 178–90.

Scanlon, Liz Garton. *All the World*. Illustrated by Marla Frazee. Beach Lane Books, 2009.

Serres, Michel. *The Incandescent*. Translated by Randolph Burks. Bloomsbury Academic, 2018.

Serres, Michel. *Malfeasance: Appropriation Through Pollution?* Translated by Anne-Marie Feenberg-Dibon. Stanford University Press, 2011.

Shakespeare, William. *As You Like It*. Edited by Barbara A. Mowat and Paul Werstine. Simon & Schuster, 2019.

Shakespeare, William. *Hamlet*. Edited by Barbara A. Mowat and Paul Werstine. Simon & Schuster, 2012.

Shubin, Neil. *Your Inner Fish: A Journey Into the 3.5-Billion-Year History of the Human Body*. Vintage Books, 2008.

Sloterdijk, Peter. *Bubbles: Spheres Volume 1: Microspherology*. Translated by Wieland Hoban. Semiotext(e), 2011.

Sloterdijk, Peter. *Foams: Spheres Volume 3: Plural Spherology*. Translated by Wieland Hoban. Semiotext(e), 2016.

Sloterdijk, Peter. *Globes: Spheres Volume 2: Macrospherology*. Translated by Wieland Hoban. Semiotext(e), 2014.

Sloterdijk, Peter. *Terror from the Air*. Translated by Amy Patton and Steve Corcoran. Semiotext(e), 2009.

Spitz, René A. *The First Year of Life: A Psychoanalytic Study of Normal and Deviant Development of Object Relations*. International Universities Press, 1965.

Steedman, Carolyn. "'Something She Called a Fever': Michelet, Derrida, and Dust (Or, in the Archives with Michelet and

Derrida)." In *Archives, Documentation, and Institutions of Social Memory: Essays from the Sawyer Seminar*, edited by Francis X. Blouin Jr. and William G. Rosenberg. University of Michigan Press, 2006.

Strand, Mark. *New Selected Poems*. Knopf, 2009.

Taylor, Astra. *Remake the World: Essays, Reflections, Rebellions*. Haymarket Books, 2021.

Turner, Christopher. *Adventures in the Orgasmatron: How the Sexual Revolution Came to America*. Farrar, Straus and Giroux, 2011.

Vanier, Catherine. *Premature Birth: The Baby, The Doctor and the Psychoanalyst*. Karnac Books, 2015.

Weiss, Hans-Rudolf. "The Method of Katharina Schroth— History, Principles and Current Development." *Scoliosis and Spinal Disorders* 6, no. 17 (2011).

Winnicott, D. W. "Birth Memories, Birth Trauma, and Anxiety." In *1946–1951*, vol. 3 of *The Collected Works of D. W. Winnicott*, edited by Lesley Caldwell and Helen Taylor Robinson. Oxford University Pres, 2016.

Winnicott, D. W. *Collected Papers: Through Paediatrics to Psycho-Analysis*. Tavistock, 1958.

Winnicott, D. W. "The Observation of Infants in a Set Situation." *International Journal of Psychoanalysis* 22 (1941): 229–49.

Winnicott, D. W. "Transitional Objects and Transitional Phenomena: A Study of the First Not-Me Possession." *International Journal of Psychoanalysis* 34, no. 2 (1953): 89–97.

Wise Brown, Margaret. *Goodnight Moon*. Illustrated by Clement Hurd. HarperCollins, 1947.

Zhang, Yu Huan, and Ken Rose. *A Brief History of Qi*. Paradigm Publications, 2001.

JAMIESON WEBSTER is a clinical
psychoanalyst, professor, and *New York Review
of Books* contributor. She is the author of
Disorganisation & Sex and *Conversion Disorder:
Listening to the Body in Psychoanalysis*.